55 Bars

55个酒吧

王姗姗 陶立军 康琪 马英伟 译

高迪国际出版有限公司 编

大连理工大学出版社

图书在版编目 (CIP) 数据

55个酒吧 / 高迪国际出版有限公司编；王姗姗等译
. —大连：大连理工大学出版社, 2013.2
 ISBN 978-7-5611-7480-7

Ⅰ. ① 5… Ⅱ. ①高… ②王… Ⅲ. ①酒吧—室内装饰设计—世界—图集 Ⅳ. ① TU247.3-64

中国版本图书馆 CIP 数据核字 (2012) 第 297843 号

出版发行：大连理工大学出版社
　　　　　（地址：大连市软件园路 80 号 邮编：116023）
印　　刷：利丰雅高印刷（深圳）有限公司
幅面尺寸：246mm × 290mm
印　　张：21.5
插　　页：4
出版时间：2013 年 2 月第 1 版
印刷时间：2013 年 2 月第 1 次印刷
策　　划：袁　斌　刘　蓉
责任编辑：裘美倩
责任校对：王秀嫒
封面设计：梁文静

ISBN 978-7-5611-7480-7
定　　价：328.00 元

电　话：0411-84708842
传　真：0411-84701466
邮　购：0411-84703636
E-mail：designbook@yahoo.cn
URL：http:// www.dutp.cn

如有质量问题请联系出版中心：（0411）84709043　84709246

PREFACE 序言

Bars, nightclubs, and lounges are playgrounds. They are favorite places of escaping, opportunities to play out our public identity, to exercise a lifestyle of entertainment, spectacle and display. These venues are designed as performance spaces, where the social dynamic of the encounter is accommodated and facilitated, where the interaction is given a tempo, where the narrative unfolds and inspires.

Their continual proliferation has compelled owners to offer evermore unique design values, which are used as a source of distinction and identity. Architects and designers have seized this opportunity to create spaces which are formally, technically, and technologically at the forefront of architectural innovation and research. The architecture of such hospitality spaces and their designs and technological systems (including lighting, sound, ventilation, materials, color, etc.) all work in unison to develop a multi-sensorial storyline in which each visitor finds his or her role. In this sense, the venue is a theater where the visitor is able to suspend the quotidian grind and indulge in a sexy, decadent experience.

This book is an indispensable compendium of outstanding entertainment venues from across the globe.

It showcases remarkable reach and technological advancements which deliver extraordinary formal and experiential values. A common item seems to be the use of computerized fabrication (i.e. laser-cutting, CNC-milling, 3D printing, etc.), which allows for endless formal opportunities and exciting environments. Notable are the varied approaches to the organization of spaces in order to cater to a varied clientele. A further recognizable trend is the tendency to create spaces which are seamless and integrated, where the structural, mechanical, lighting, and sound systems perform without overbearing the purely formal design systems.

The projects in the following pages portend an ever-evolving scenario of exciting developments.

Enjoy this parade of extraordinary venues where visitors quench their desire to be on stage and reach a new exciting state, full of anticipation.

Who knows what the night might hold in store?

Antonio Di Oronzo, principal
bluarch architecture + interiors + lighting
Professor of Architecture (City College of New York—Spitzer School of Architecture)

  酒吧、夜店和休闲吧这些极佳的休闲场所给人们一个逃离现实、摆脱社会身份的机会，体验不同凡响、尽情释放的精彩生活方式。这里是展示自我的舞台，上演着人生中种种令人悸动的邂逅，这里有着富有节奏感的沟通和令人唏嘘的传奇故事。

  新的酒吧不断涌现，店主们必须追寻更为独特的设计理念，以作为一种与众不同的身份标识。建筑师和设计师们把握住这一机会，创造出在形式、工艺、科技上都处在建筑业创新和研究前沿的空间。这些娱乐场所的建筑构造、设计理念和所用技术（照明、音响、通风、原材料、色彩等）和谐地展现出多视角的故事脉络，每位身处其中的客人都能找到自己的角色。从这种意义上说，酒吧就像是一间戏院，客人在此可暂时忘却现实中的琐事，纵情于纸醉金迷的片刻。

  本书为全球最佳娱乐场所的必备指南。

  本书展示了非凡的视觉极限和先进科技，传达出酒吧不同寻常的结构和经验价值。众多案例设计的共同之处大概在于计算机操控设备（例如，激光切割、数控铣床、三维立体刻印等）的运用，使不计其数的外形创意和精妙的环境设计成为现实。值得一提的还有设计师按照客户所需以相应的方式去规划空间。另外一种趋势是力求打造出浑然天成的空间：在这个纯粹的设计系统里，结构、机械、照明、音响系统都不会喧宾夺主。

  本书案例向读者展示了一幅幅不断演变的、激动人心的酒吧设计场景。

  读者可带着期许，在书中尽情欣赏这些令人惊叹的设计：客人在此满足了展示自我的渴望、获得更兴奋的心情。

  谁能猜到在午夜的酒吧里会发生怎样的奇遇呢？

安东尼奥•迪•奥伦佐
布拉齐建筑+内部装饰+灯光照明设计公司
纽约城市大学斯必泽建筑学院建筑系教授

When speaking about bar design, Lionel Ohayon, the founder of ICRAVE a New York-based hospitality design firm, says that it is best approached from a sociological standpoint: "we understand and think of each design element in a bar in terms of urban planning, cities, and how people interact with each other on a larger scale. We bring that larger vision down to the very minute details of a bar, always keeping focused on interaction." The goal in bar design is to create a venue where guests are encouraged to relinquish their inhibitions and step through the looking glass to experience an alternate reality for the night.

The atmosphere at ICRAVE-designed bars is curated using state-of-the-art A/V systems that seduce patrons while they dance the night away.

To create a more intimate bar, we often design a central space and tiered seating lay out allowing for a see-and-be-seen environment where the energy is focused to the middle of the room. This allows the space to flow in such a way that patrons have multiple options for discovery and interaction. The ICRAVE design team always pushes the envelope on inventing spaces that allow individuals to experience explore and more importantly become a part of the environments they are visiting.

ICRAVE

当谈及酒吧设计，Lionel Ohayon——总部位于纽约的ICRAVE建筑设计公司奠基人——称从社会学角度考虑才是理解设计的最佳手段："我们从城市规划、城市特点和大批人群彼此互动的方式等方面了解并思考酒吧内每一个设计元素。我们把较大的视角细化到酒吧的每一个微小细节，时刻将人与人之间的互动当作焦点。"设计酒吧的目的是为了创建一个场所，鼓励人们在此释放自己，不止于驻足观望，而是走进去体验夜晚另外一种生活。

ICRAVE设计的酒吧氛围均由世界最先进的音响效果控制，当客人们整晚跳舞之时，营造一种充满诱惑力的氛围。为实现外观与功能的完美结合，酒吧的设计之道要求其每一寸空间都可成为舞台。

为使酒吧更具私密性，我们经常设计一个中央空间和阶梯形坐席区，这种观赏-表演区结合的布局可以让众人将注意力都集中到房间中央。该通透的空间设计也使客人们在发现新奇之处和交际互动时有多种选择。ICRAVE设计团队经常在空间设计方面推陈出新，尤其擅长营造让客人拥有充分体验、并融于环境的空间。

ICRAVE设计公司

CONTENTS 目录

Nuts 4	**008**	**112**	Velvet Bar
Nuts 5	**016**	**116**	The Carbon Bar
Chan	**024**	**122**	Cirus
CIRCA	**032**	**128**	Meltino Bar & Lounge
Club Bach	**038**	**136**	Oxo Tower Bar
MIXX BAR & LOUNGE	**046**	**144**	Press Club Urban Wine Venue
Bar Kottulinsky	**052**	**150**	Na Mata
Lounge MS	**058**	**158**	Hyde Lounge
Bar Rouge	**066**	**164**	Cienna Ultralounge
Waves Bar	**074**	**168**	Dusk
YUCCA	**078**	**174**	Tamarai Lounge & Restaurant
Zebar	**084**	**182**	VYNE
Plan B	**092**	**186**	Numero Bar
Brown Sugar Taipei	**096**	**192**	La Folie
Pinon Grill	**104**	**196**	Library Bar

Monkey Bar Fumoir	**202** ◆
Narcissus Bar Restaurant	**208** ◆
Obikà Mozzarella Bar	**212** ◆
Soe	**216** ◆
26 Lounge Bar	**222** ◆
Q Bar	**226** ◆
Simyone Lounge	**230** ◆
Villa	**234** ◆
XVI Lounge	**240** ◆
Starlite	**246** ◆
Mouton Cadet Wine Bar	**252** ◆
Buck and Breck	**258** ◆
Eighty-Six	**262** ◆
Asphalt	**268** ◆
Tribeca Grand Hotel Lounge	**276** ◆
Fashion Bar	**280** ◆
The Mansion	**284** ◆
W Lounge Bar	**290** ◆
WYLD Bar	**294** ◆
Republic Gastropub	**298** ◆
Bar Fou Fou	**304** ◆
BAIXA	**310** ◆
MUGEN	**316** ◆
Twenty Five Lusk	**322** ◆
Disko Bar Mladost	**332** ◆
INDEX	**336** ◆

Nuts 4

Designer: Jorge Luis Hernández Silva
Design Company: Hernandez Silva Architects
Location: Guadalajara, Jalisco, Mexico
Area: 750 m²
Photographer: Carlos Díaz Corona

The concept born from the entrance, the facade is a large screen and its base has irregular seats that are connected to the reception through a tunnel of arrows, a fully isolated box of enclosure strips on floor and ceiling. Inside the juxtaposition of old structures serve as the motive of the project, plans continued to arise from the floor and wrap up the walls and ceilings of various sizes. A huge bar that winds through the walls integrating the three bars as one, using the bottles as monitors, reflecting the textures, colors and lights. The track is sunk as if it was a theater. The niches located next to the track reinterpret a railroad car. The project takes up forgotten items and materials, giving a new expression and incorporating the project as an experiment: velvets, tergal, glass mirror, carpeted walls, glass spheres, curves, etc. are some of the materials used to emphasize the concept. The colors are now of completely lively.

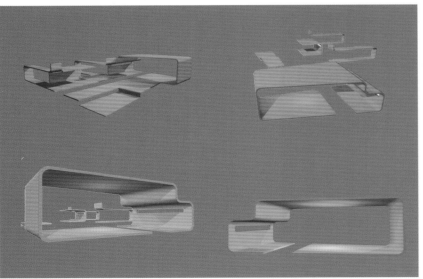
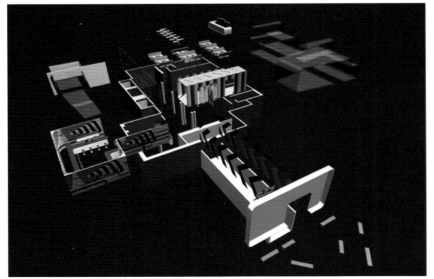

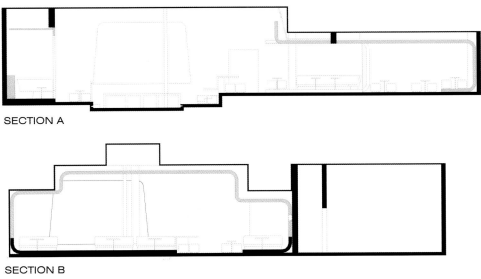

SECTION A

SECTION B

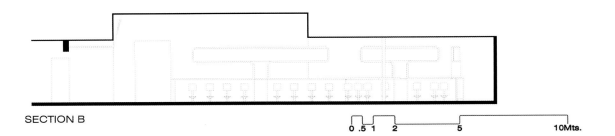

SECTION B

0 .5 1 2   5   10 Mts.

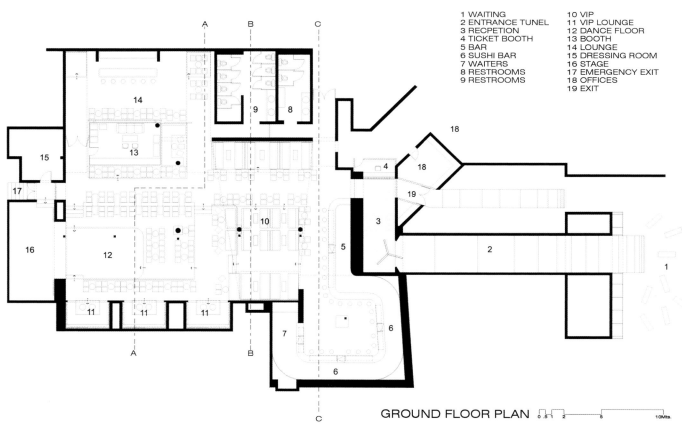

1 WAITING	10 VIP
2 ENTRANCE TUNEL	11 VIP LOUNGE
3 RECPETION	12 DANCE FLOOR
4 TICKET BOOTH	13 BOOTH
5 BAR	14 LOUNGE
6 SUSHI BAR	15 DRESSING ROOM
7 WAITERS	16 STAGE
8 RESTROOMS	17 EMERGENCY EXIT
9 RESTROOMS	18 OFFICES
	19 EXIT

GROUND FLOOR PLAN

本案设计理念源于入口处,其外立面是一个大屏幕,底层放置了很多不规则座位。穿过画满箭头的通道,一直通往前台,便进入到一个完全独立的包厢,天花板和地板被带状图案所围绕。设计主旨体现在内部并排放置的古旧设施中,图案从地板一直向上延伸,包裹住不同面积的墙壁和天花板。一个巨大的吧台蜿蜒穿过墙壁,令三个酒吧间合而为一,并用酒瓶来映射出质地、色彩和灯光。酒吧内的过道像电影院里的一样拾阶而下。设置在过道旁的壁龛重新诠释了火车车厢的形象。本案把那些被人遗忘的物品和材料纳入设计中,并给予其全新的表达,就像在做试验:天鹅绒、泰格尔纤维、玻璃镜面、软包的墙壁、玻璃球、曲线等等,都被用做强调设计理念的材料。酒吧的色彩非常鲜活。

Nuts 5

Designer: Jorge Luis Hernández Silva
Design Company: Hernandez Silva Architects
Location: Guadalajara, Jalisco, Mexico
Area: 750 m²
Photographer: Carlos Díaz Corona

The theme lies in the exaltation of the irregularities of the existing form, that are intertwined in multiple directions as an irregularly folded paper that is wrapped around a space, born from the outside with a box that floats and throws against access by generating the entrance, as a deep red envelope. The intensity is turned off when passing through the tunnel. Everything becomes completely gentle and pure white when entering. These docile and pure lines gradually transform themselves and break to finish within a chaos of enveloping grey and red. Everything here is exciting, and there are no walls or ceilings but a broken cover and bright and segmented images. The cabin spaces, walls and furniture are in a powerful, bright red, and the walls are segmented images that emphasize the concept.

本案的主题是试图将现有建筑的不规则形状加以升华，这些不规则的形状互相交织，朝不同方向伸展，就像一张折叠不规则的纸，将一个空间包围起来，由外而内形成一个浮于空中，正对入口的盒状结构。入口是一个深红色外壳。自外而内进入时，一切变得柔和而洁白。这些柔和而洁白的线条逐渐变化，在一片红灰色调中戛然而止。这里的一切都令人兴奋。没有围墙，没有天花板，只是一层呈碎裂状的覆盖物与被分割的图像。包间、墙壁和家具均是具有冲击感的亮红色。墙壁也是镶块式构图，强调了设计的主题。

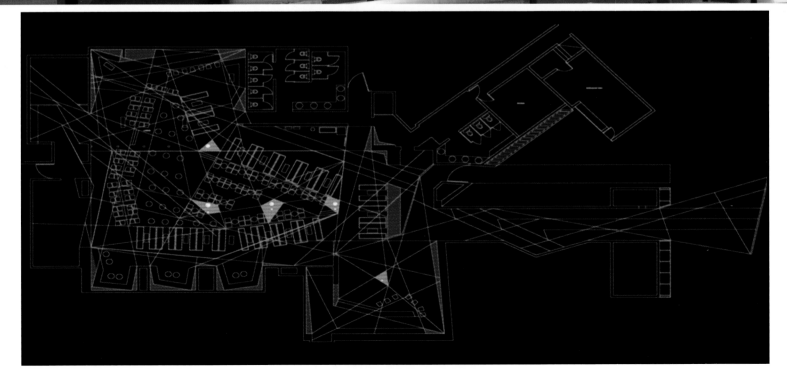

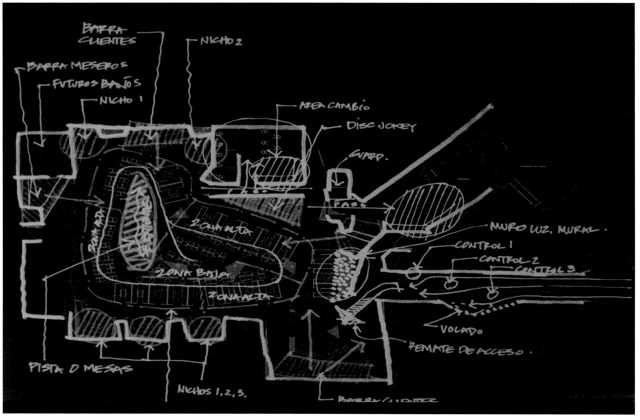

Chan

Designers: Andy Martin, Art Waewsawangwong, Daniel Rodriguez, Tom Davies
Design Company: Andy Martin Architects
Graphic Design: Farrow
Carpet Design: Yulia Bakhtiozina
Location: Thessaloniki, Greece
Area: 230 m² (internal) and 150 m² (external)
Photographer: Vangelis Paterakis

Chan won "best lighting award" at the prestigious Restaurant and Bar Awards 2011. AMA design studio's concept for Chan "pan-Asian" restaurant and bar at "The Met" hotel, is a lively fusion of modern aesthetic and traditional Asian motifs. Designed by Australian architect, Andy Martin, it opened to acclaim for its dazzling interior and innovative cuisine, Chan is the "destination".

AMA extensively researched their brief from "The Met" with trips to China Hong Kong, Thailand and Indonesia, trying to seek out the essence of distinctly modern Asian cuisine. The interior is conceived as a dynamic masculine space, with its furniture tactile and feminine. The seating is arranged in a central sculptural element which floats within the charcoal anodized box. Intricately crafted and luxurious semi-private seating makes Chan cosy and intimate. The seating booths cater for two or for larger groups. There lies a dramatic sense of layering within the space from the graphically lit wall panels, open grid ceilings, to the engraved detailing on the tables. Tattooed graphics on upholstery and 21st century "manga" interpretations within the cocktail lounge; this becomes an experience of Asian street culture and cosmopolitan sophistication.

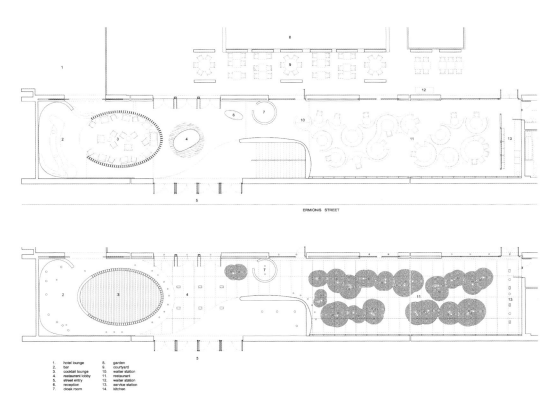

1. hotel lounge
2. bar
3. cocktail lounge
4. restaurant lobby
5. street entry
6. reception
7. cloak room
8. garden
9. courtyard
10. waiter station
11. restaurant
12. waiter station
13. service station
14. kitchen

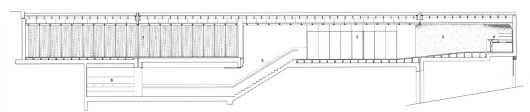
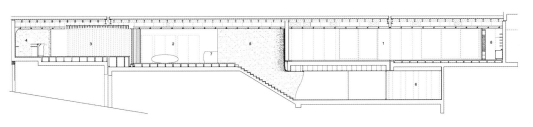

1. restaurant
2. restaurant lobby
3. cocktail lounge
4. bar
5. stair
6. toilet
7. reception
8. service station

  2011年在餐厅酒吧界享有盛誉的设计奖项角逐中，幽暗酒吧赢得了最佳灯光设计奖。AMA设计工作室在幽暗酒吧"泛亚"餐厅和"遇见"酒店中酒吧的设计理念上实现了现代美学元素和传统亚洲图案的完美结合。在澳大利亚建筑师安迪·马丁的设计下，幽暗酒吧彰显出眩目的内部设计和独具匠心的菜肴，使它成为不可错过的目的地。

  AMA设计工作室通过走访中国香港、泰国和印尼等地，对"遇见"酒店的概况进行了广泛深入的研究，企图发掘出现代亚洲烹饪的精髓。本案内部被构想成一个动态的粗犷空间，但在家具上则设计得更具触感，更柔和一些。座椅的排列借助中央雕纹元素理论，似在深灰色光亮的盒子中自由漂移。精致奢华的半私密坐席让酒吧散发出一种惬意和亲切的气息。包厢适于两人或多人使用。从绘画装饰的墙板、开放式的网格天花板，到餐桌上细致的雕刻，无处不透射出一种动态的层次感。室内装潢的平面图案和鸡尾酒餐厅内对21世纪漫画的诠释，都为身处幽暗酒吧的客人提供了一次体验亚洲街头文化和大都市复杂生活的经历。

CIRCA

Design Company: 3six0 Architecture
Location: Memphis, Tennessee, USA
Client: John Bragg
Area: 398 m²
Photographer: John Horner

The screens frequently seen on building facades in Memphis served as the inspiration for this design. While these facade screens function to soften the strong sunlight, the designers employed layers of screen to address the client's need for "the mayor to have lunch without anyone noticing" while preserving the open face running along a longitudinal side. The screens' patterning morphs to accommodate their function as a wine wall, light wall or curtain. As one passes in between layers of screens or alongside the glazed long side, a moiré pattern appears. It literally looks like water.

The screens create layered areas of space while maintaining an overall sense of openness. Three different screens were created to address different programmatic needs. All screens were digitally cut from the architects files and fabricated offsite.

Etched finish film cut into squiggles have been applied to the 38.1 meters long glass storefront facing the pedestrian arcade. The widths and spacing of the squiggles are varied to create different degrees of privacy. This squiggle is the fundamental component of the screen geometry.

The layered screens of aluminum squiggles and cherry veneer panels separate the dining areas from the restaurant's main circulation. The layered screens have the optical phenomenon of a moiré when one passes by, but visually appear open when viewed directly. The screens double as a wines wall which allows the chef to display his full collection of wines.

The rear cherry paneled wall is a graphic interpretation of the screen wall's moiré effect. The layers of the screen have been collapsed into a two dimensional surface that is a moment of the moiré. The paneled wall is backlit from behind and provides a large portion of the dining area and bar lighting.

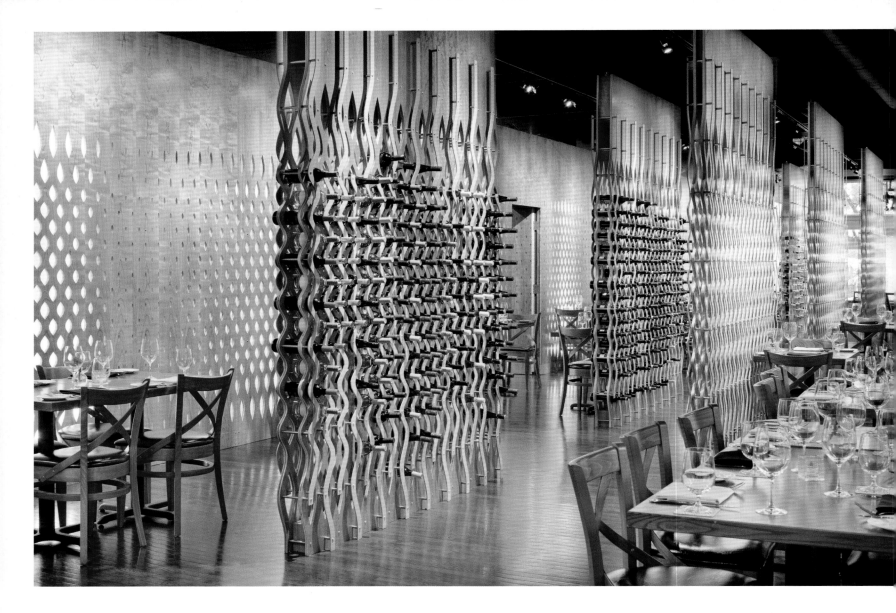
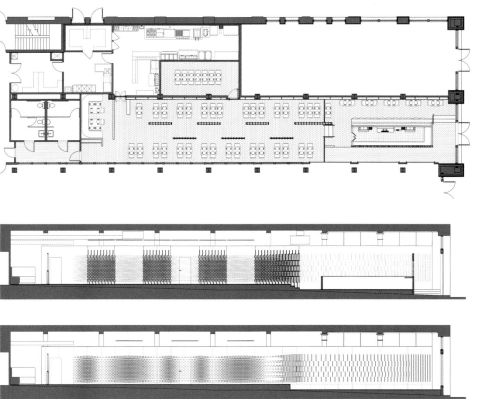
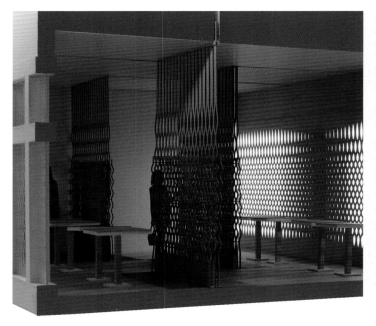

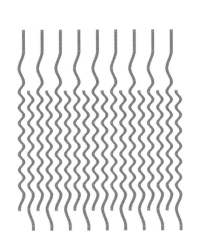
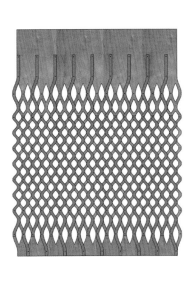

在孟菲斯城，建筑物的正面外墙经常能看见格网，这是CIRCA的设计灵感。虽然这些外墙格网的功能是柔化强烈的阳光，然而设计师们运用多层格网来满足客户"连市长都能够隐秘地用餐"之需求，同时保留纵向延伸的开放式表面。格网的图案结构不断变化，以适应其作为葡萄酒墙、光墙或帷幕的功能。当一个人走到格网的各层之间或沿着玻璃长边行走时，便能看到波纹图案，看起来像水一样。

格网制造出分层的空间，同时又保持了一种开放的整体感。三种不同的格网可以满足不同的可控需求。全部格网都按照建筑师的设计数控切割而成，并在场外组装。

蚀刻表面的薄膜被分割成波形曲线，安装在面向行人拱廊的38.1米长的玻璃店面上。

波形曲线的宽度和间距不断变化，制造出不同程度的隐秘感。波形曲线是格网几何结构的基本组成部分。

铝制波形曲线和樱桃木制面板的分层格网结构把用餐区和餐厅的主流通区分隔开来。当有人经过时，分层格网会显现出波纹的光学现象，但直接看去又显得开阔。双层格网用作葡萄酒墙，主厨可在此陈列葡萄酒藏品。

后部的樱桃木面板墙是格网波纹效果的图形诠释。将各层格网通过二维平面表达出来，展现了波纹效果的一个瞬间。镶以嵌板的墙壁采用背光，为用餐区和吧台提供了大部分照明。

Club Bach

Designer: Yasumichi Morita
Design Company: GLAMOROUS co.,ltd.
Location: Osaka, Japan
Area: 166 m²
Photographer: Nacasa & Partners Inc.

Club Bach, located at a prominent commercial avenue in Osaka, has been designed to be the most luxurious lounge in the world and many luxurious surprises await you in each area from the elevator hall to the reception and the main hall.

In the elevator hall, the beautiful door with layered fabrics enclosed in the glass produces a sophisticated atmosphere. When you step into the reception area from the elevator hall, you will be offered extraordinary experiences: the chandeliers seems like endless because of the half mirror effect while the dark mirror vanishes the presence of the wall. In the floor, you will see the gorgeous portrait artwork by Mr. Joe Satake lies down. The space fuels great curiosity among guests, like a private gallery of Roman.

In the main hall, the walls are lavishly covered with special panels of stainless and some hundred-thousand crystals, featuring bronze mirrors at random. In the middle of the space, there is a brilliant horse object studded with crystals for a playful touch. Despite the solid material composition of mirror, stainless and crystals, thanks to the bronze coloration and warm indirect lightings which create various expressions of lights and shadows, the air is not only elegant and also relaxing.

Every material and light fixture was carefully selected especially to enhance the beauty of ladies' expressions. For example, the combination of white marble of the tables and down lights produces soft halation which is good for adding a glow to the skin.

In the VIP room, there are mirrors at four corners to open up the space visually and the dynamic objects made with pink gold beads make the space rhythmical and impressive.

Even if you have seen the best things in the world, you will still be amazed by this great flamboyance.

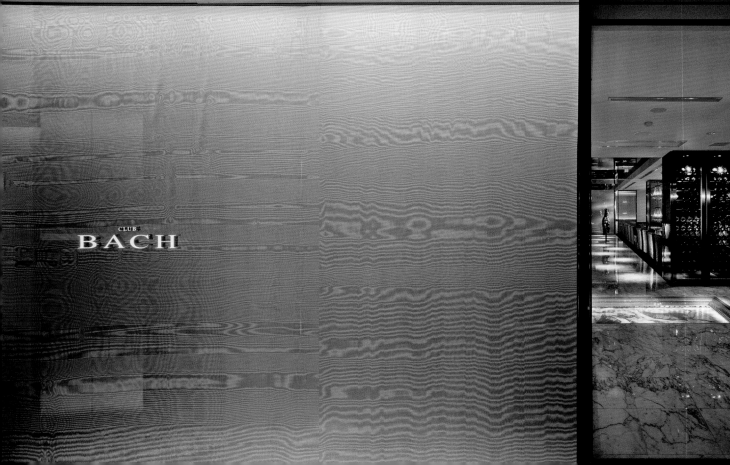

巴赫酒吧位于大阪顶级商业街，被设计成世界顶级豪华休闲酒吧，从电梯大厅到前台乃至主厅，数不尽的奢华精巧的设计使客人大开眼界。

在电梯大厅美丽的门上，封装于玻璃之中的层层叠叠的纺织品营造出一种精致的氛围。当客人从电梯大厅步入前台区域时，将会有与众不同的体验：由于半镜面效果，枝形吊灯好像延绵不尽，而墨色玻璃从视觉上掩藏了墙体的存在。在地板上能看到乔·佐竹所画的华丽人物像。整个空间好似古罗马私人画廊一般，使客人对其好奇心剧增。

在大厅内，墙体大面积地被特制不锈钢板材和数十万块的水晶覆盖，随机地饰有几面铜镜。空间中央摆放着一座由水晶装饰的马形摆设供大家触摸把玩。尽管空间主要由镜子、不锈钢、水晶灯等硬朗材质装饰，但由于青铜色调和暖色间接照明折射出的各种光影效果，使空间氛围既优雅又令人放松。

所有的材料及灯饰都经过精挑细选，以期为女士的姣容再增一份美感。比如，白色大理石的桌子与下射灯营造出一种柔和的光晕，可以将皮肤映衬得更加有光泽。

在贵宾厅，四角都饰有镜子，从视觉上扩大空间。粉金色珠子穿成的装饰物使空间更有韵味和张力。

即便已经见识过世界上最好的东西，但仍会对这里的奢华感到震撼。

MIXX BAR & LOUNGE

Designer: Gwenael Nicolas
Design Company: Curiosity inc.
Location: Minato-Ku, Tokyo, Japan
Area: 600 m²

Spanning 600 square metres at the top of the 36-floor hotel, MIXX (at the Intercontinental Hotel) is a playful exploration of light and shadows offering an atmospheric window over the city of Tokyo.
The entrance is marked by a "gate of light": a wall of floating white fabric sculptures reflected through a play of mirrors, their delicacy balanced against the dynamism of the lighting.
Covering the floors is a uniquely designed carpet whose hues and patterns evoke the natural outdoor moss formations of a traditional Japanese garden.
The first of three areas is an open bar space – complete with a randomly placed high counter, a warm and sociable spot for friends and dates meeting for cocktails or pre-dinner aperitifs before visiting the Pierre Gagnaire restaurant on the same floor.
Next is the main bar area. A signature 10 metre long counter sets a bold tone, taking centre stage against a backdrop of windows that reveal Mount Fuji by day and Tokyo's digital light show by night.
Finally, there is the lounge. A white artwork of fabric hovering in the dark marks entrance, leading to a central table and more intimate sofas tucked snugly into the zig-zag line of windows that surround the space. Cocooned by a series of vertical lights, the lighting dynamically cuts through the atmospheric dark space.

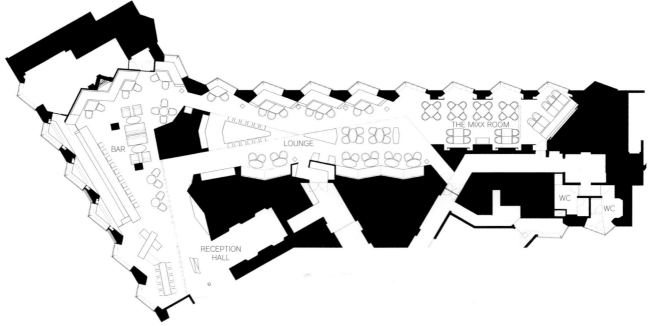

洲际酒店内的迷思酒吧占地600平方米，位于酒店顶层36层，室内设计以有趣的手法探索光与影之间的关系，并为顾客提供鸟瞰东京的绝佳视野。

入口标志是一扇"光之门"：一排浮动的白色织物的雕塑，像一面墙映射于镜中，设计兼顾了雕塑的精致与光影的活力。

本案地面铺着独特设计的地毯，其色彩和图案都唤起人们对传统日本庭园的自然户外苔藓的形态的记忆。

本案分为三个区域，第一个区域是开放式酒吧，随意放置着吧台，营造出热情的社交空间。朋友们可相约于此喝鸡尾酒，或是在去同一层楼的皮耶·加尼叶餐厅之前，来这喝杯餐前开胃酒。

接下来是主酒吧区。一个饶有特色的10米长桌在中心位置奠定了醒目的基调，白天凭窗可观赏富士山，夜晚可欣赏东京的数码灯光秀。

最后一个区域是休闲吧。入口暗处显现的是一件白色编织艺术品，引导客人到主台。四围曲折的窗边贴合地安置着舒适的沙发。休闲吧由一系列垂直的灯具所包围，灯光动态地穿透了这富于艺术气息的幽暗空间。

Bar Kottulinsky

Design Company: LOVE architecture & urbanism
Location: Graz, Austria
Area: 350 m²
Photographer: Schuller Jasmin

The bar was divided into three core areas, which comply with the new non-smoking protection laws:
- The Warm-Up Zone: The front part of the Kottulinsky (near the main entrance) features a centrally aligned bar. This area offers a relaxing place for guests to chat in the early evening hours (non-smoking area).
- The Dance-Floor Area: This is the dancing area, which is festive and loud (smoking area).
- The Lounge: The back area of the Kottulinsky comprises a lounge with an intimate, communicative and quiet atmosphere (non-smoking). The lounge's mobile furniture provides plenty of freedom to arrange the seating and lounging area in different ways.

Smooth, glass partitioning walls divide these three areas both climatically and acoustically.

Stylistically, the visible brick wall of the vaulted cellar remains. The colors, materials and surfaces of the furniture and fixtures complement the brick walls: leather, wood, stainless steel and historic wall coverings are the predominant materials.

Indirect, ground LED spotlights provide the basic lighting, which illuminates the barrel vault and creates a subtle play between light and shadow along the visible brick wall. Moving heads are installed on the transverse arches in the dance-floor area, which lend a rhythm to the room structure. The Lounge area is illuminated with projectors, which transforms the Lounge into a spherical experience.

Light frisbees are the highlight of the illumination concept. They are LED lights that appear to hover in the room and that – depending on the setting – shine in different colors and light designs. The light frisbees were loosely distributed with different diameters. They suffuse the various areas of the bar in different ambiences of light and color.

The "Permanent Unit" street artists created large-scale interpretive structures that are distributed freely throughout the vaults. They give the room a kind of "new perspective layer": the free formations make the vault appear larger or shorter, broader or narrower, depending on the perspective. At the same time, the viewer constantly notices new abstract-realistic formations or sketches.

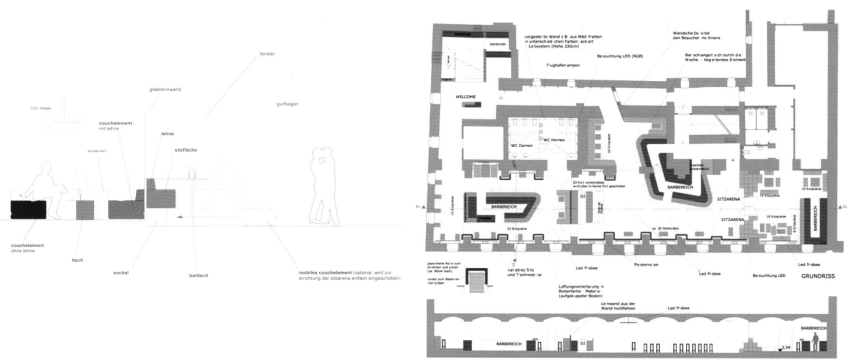

克图林斯基酒吧分成三个核心区域，这样的设计符合当地新的非吸烟者保护法。

-交友区：克图林斯基酒吧的前部（位于正门入口处）有一个吧台。该区域（非吸烟区）为客人提供了一个傍晚轻松闲谈的场所。

-舞池：装修华丽、充满欢乐气氛，客人可在此翩翩起舞。

-休息区：位于克图林斯基的后部。气氛亲切幽雅，适于畅谈交流（非吸烟区）。该区的家具可以自由挪动，以不同方式满足休闲需求。

光滑的玻璃幕墙把三个区域分隔开来，以营造不同的气氛，满足各自的声效需求。

在风格上，拱形酒窖的外露墙砖得以保留。家具的色调、材质、外观效果和墙上的饰品与砖墙的效果相得益彰。这些饰品的主要材料是皮质、木质、不锈钢和具有沧桑感的墙面包覆材料。

间接照明的LED（发光二极管）地装射灯为酒吧提供基本照明，既可照亮半圆形的拱顶，又沿砖壁营造出微妙的光影效果。在舞池区域的横向拱顶上安装有摇头灯，为室内结构平添了律动之感。休息区由投影仪照明，使人感觉仿佛置身一个球形空间。

在照明设计的理念中，飞盘形照明灯是个亮点。这些用来照明的LED灯具好似在房间上空盘旋。根据背景的不同，它们仿佛呈现不同的颜色。这些飞盘状灯具分布随意、大小各异。在酒吧的各个区域，它们渲染了不同的光色氛围。

街头艺术家们在各个拱顶部分都大规模地采用了诠释结构模式，这些结构使得房间拥有了一种"新透视层次"：这些自由的结构使拱顶显得或大或小、或宽或窄，完全取决于透视角度。同时，观察者总能发现或抽象或具体的新的格局或设计手绘。

Lounge MS

Designers: Antonio Vaillo, Juan Luis Irigaray, Daniel Galar Irurre
Design Company: Vaillo + Irigaray Arquitectos
Photographer: Jose Manuel Cutillas

The organizational scheme is due to similar patterns of micro-structures, more in line with geometric patterns of liquid and / or aerosols that Cartesian structures. In establishing a working geometry using "soft" and a unified treatment space "airy".

The new space is conceived as a continuation of the existing fence, wrapping it all, but hidden: expressing their new identity, not as constructed element, but as a re-forested. A new plant species grown in the surrounding area, the new scalar similarity between elements of "plant" makes the proposal a new understanding of the link with the existing connection.

This species builds a base of recycled plastic tubes of different colors similar to the natural, which like "reeds" is organizing a braided flexible and deformable organic capability to adapt to any situation and geometry.

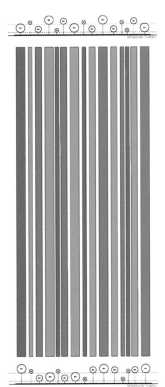
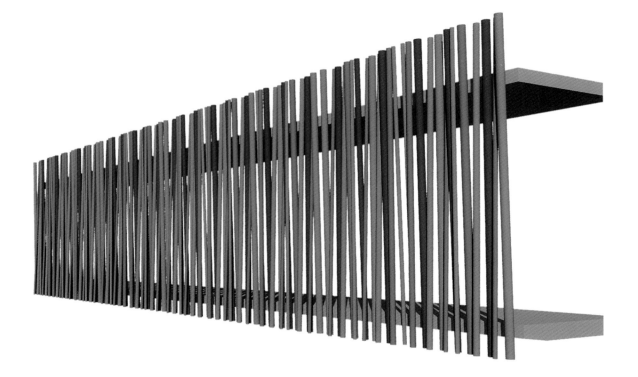

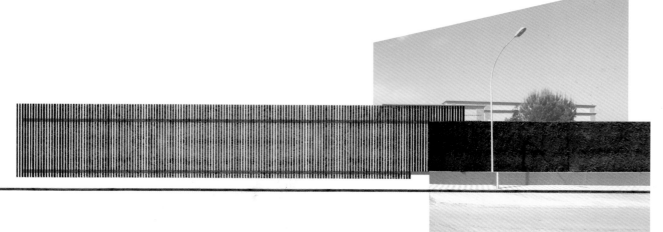

本案设计方案源于与之相似的微结构样式，比较符合液体几何构造或者笛卡尔提出的气胶体结构。在建造这种几何结构时，使用了统一的空间柔化手法，显得非常灵动。

新建的酒吧空间被构想为现有酒吧空间的延续，环绕整间酒吧，却毫不突兀：它将自身标识为原有空间的延伸而非新加入的元素。新品种的植物被栽种在周围。所采用植物元素在数量上的相似性，使得本案成为对新旧空间关联性解读的一个新范例。

这些植物为那些颜色各异、形似芦苇的可再生塑料管构建了基础，这些塑料管柔软可编织，能够做出各种几何造型来装饰不同的空间。

Bar Rouge

Team: Selene Lu, Marc Antoine Melia, Jay Louin, Mathieu Brauer, Antoine Khan, Michel Wilkins, Karina Palomin
Design Company: Naço Architectures
Location: Shanghai, China
Area: 1000 m²
Photographers: Vincent Fillon, Tristan Chapuis

Naço Architectures recently re-design the new Bar Rouge, one of Shanghai's most legendary nightspots.

"Two nights in one" is the inspirational mantra of designers — enable party-goers to enjoy a relaxing atmosphere in the early evening and a high energy club environment after 11 P.M. Naço Architectures has therefore transformed Bar Rouge by opening its interior, making it more elegant and enjoyable as a whole.

Designers conceive the new space to glitter like a jewelry box. The bar at the heart of the room locates beneath a red structure and invokes the image of a sparkling ruby when illuminated. The luminous, red netted structure is draped from the ceiling, walls and hangs above VIP areas. The DJ booth, with its faceted angles, recalls the shape and luminosity of a precious gem. Around the main bar, new VIP sections partition the space, the triumph of these being the Crystal VIP Room. This dazzling all-white exclusive space overlooks the whole bar, enabling VIPs to relax and watch the lounge transform into a night club as the night rolls on.

As the breezy indoor/outdoor architectural spaces on Ibiza, the party has no limits. A glass partition has been created between the interior and terrace to look as though it's fading away, as if by magic. The terrace utilizes a color scheme of black and gold to create an elegant party atmosphere. Here, Naço Architectures again gives the bar a central position, allowing energies to circulate, like a rock in fusion.

On Bar Rouge's terrace, whether seated in a comfy armchair, danced the night away, or curled up in the Crystal VIP room, one will find himself in the epicenter of modern Shanghai, where the sky and the lights of Pudong mingle magically.

　　纳索建筑事务所的设计师近期为新开业的红唇酒吧进行了设计，这可是上海最富传奇色彩的夜场之一。

　　"一晚两夜"的概念是设计师的灵感来源。吧迷们可在早夜场享受轻松的时光，也可在午夜时分点燃激情。纳索的设计师们为此将内部空间全部打通，让人享受大空间的优雅和惬意。

　　在设计师的构思中，新的吧场要如珠宝箱一般异彩纷呈：坐踞中央的吧台位于一个红色结构的下方。照明开启时，使人联想到光彩夺目的红宝石。网状发光的结构自天顶和四壁垂下，悬于贵宾区上方。主持人控制室设计成琢面形状，使人立刻想到璀璨的宝石。主吧台周围是一个个新设计的独立的贵宾区域，最为典型的就是水晶贵宾房。这是一个令人炫目的全白独立空间，可俯瞰整个酒吧。贵宾们可以在此小憩并在夜幕降临时看到酒吧变身夜店的全过程。

　　伊维萨岛风格的微风习习的室内外建筑空间，使这里的聚会毫无局促感。酒吧内部空间和露台之间的玻璃隔断设计使酒吧给人渐退之感，如魔术般神奇。露台使用金色与黑色的色彩搭配，创造出优雅的聚会气氛。在这里，纳索的设计师再次赋予酒吧以中心地位，让激情如熔化的岩石般在人群中传播。

　　身处酒吧的露台，无论是憩于舒适的扶手椅上，还是在深夜热舞，亦或是在水晶贵宾房蜷身而坐，你都会发现自己正置身于现代的沪上中心，欣赏着浦东繁星般的灯光融于夜空的幻景。

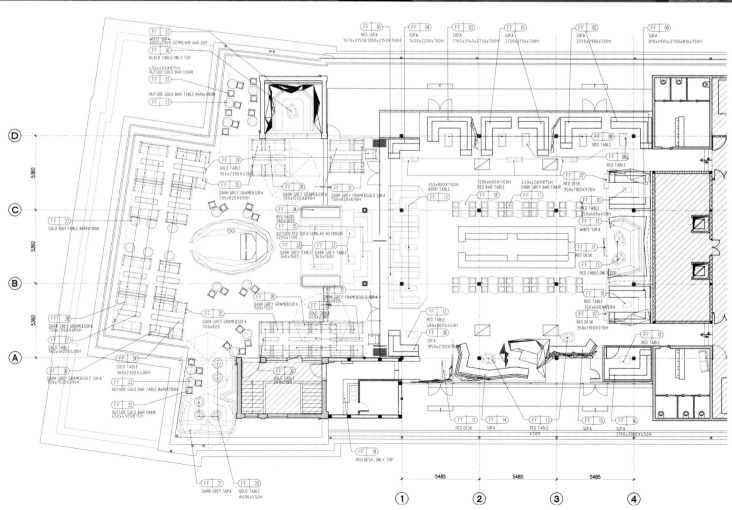

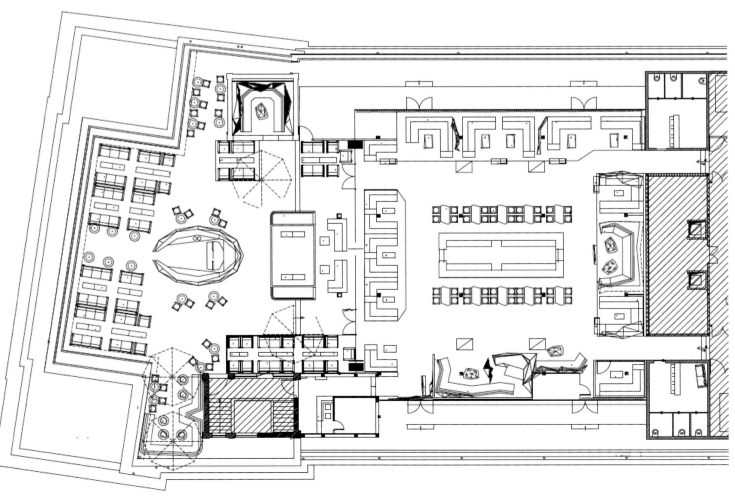

Waves Bar

Designers: Kalhan Mattoo, Santha Gour Mattoo, Hina Chudasama, Prashanta Ghosh, Gauri Argade
Design Company: Planet 3 Studios
Location: Thane, India
Area: 140 m²
Photographers: Mrigank Sharma, India Sutra

The ceiling with its undulating waves of varying pitch evokes the surface of water in mild agitation. A tiled pattern cast and handcrafted in plaster, has four distinct modules that repeat geometrically to blanket the entire ceiling and one feature wall. Painted white, it scatters light from a recessed cove along its periphery. The two sides of the space that open to the outside have a carefully articulated skin layered in glass and thermoformed Plexiglas panels. While the outer glass skin is frosted, it reveals the outside through transparent strips that follow the form of panels inside. These panels anchored to ceiling and floor twist and turn offering a subtle suggestion of a water cascade frozen in time. The effect is highlighted by a strip of programmable color changing LED lights that illuminate the cavity. The bar counter is a seamless construct clad in solid-surface acrylic. At the front, a water trough is revealed through round punctures in the apron. Air bubbles continually form and dissipate in this trough and lighting effects lend sparkle to the fizz. The bar backdrop is formed of ribbon like panels that flow along the wall shaping shelves for display of liquor bottles. Fluid lines find their way into the central seating area in form of under lighting below a frosted glass floor.

　　本案天花板犹如起伏的波浪激起整个水面。由石膏打制、手工制成的图案包括四种不同的模块,呈几何状重复排列,覆盖了整个天花板和一面装饰墙。白色涂漆使得光线可以沿着图案边缘的凹槽散射开去。空间朝外的两侧是用玻璃和热成型树脂玻璃板精心拼接的分层外立面。尽管外层是磨砂玻璃,但依然可以透过与内层嵌板图案一致的透明带看到外面。这些嵌板固定在天花板和地面之间,旋曲的造型令人联想到被冻结住的瀑布。设计还使用了一排由程序控制变色的、照亮内部的LED灯来突出这一效果。吧台是一个丙烯酸树脂漆涂层固面的无缝结构体。在吧台前面,通过台口的圆孔可看见一个水槽。气泡在水槽中,随着不同的灯光效果不断地形成、消失。酒吧的背景由类似缎带的嵌板构成,沿着与墙壁形状一致的酒架延展。磨砂玻璃地板下透出的光亮流动线条一直蔓延到中央坐席区。

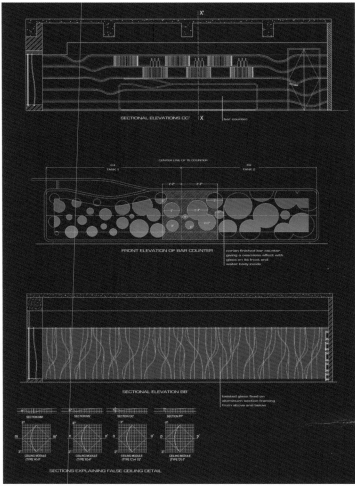
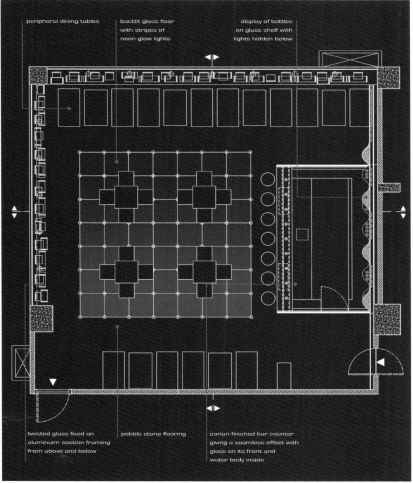

YUCCA

Designers: Thomas DARIEL, Benoit ARFEUILLERE
Design Company: Dariel & Arfeuillere — A Lime 388 Company
Location: Shanghai, China

Yucca is located at Sinan Mansions 26F, an eye-catching traditional-style building in the old French concession of Shanghai. The building is the headquarters of Yucca's creator, the famous Australian-Greek chef David Laris, and it also houses three of his other restaurants: The Fat Olive — a Greek style restaurant, 12 Chairs — a new high-class restaurant, and The Funky Chicken — a fashionable fast food. Yucca is featured by a modern atmosphere and the eye-catching design and decoration.

When you relate the word Mexico to restaurants and bars, a fixed routine will be brought to your mind: cacti, sombreros, holsters, bandoliers, cracked stucco exposing faux adobe bricks or the Frida Kahlo portrait for those places attempting to "class it up". However, the creative designers of Yucca — Thomas Dariel and Benoit Arfeuillere — did not want any of that. Far from being folkloric, Yucca pays homage to the rich visual cultures of Latin America and the Iberian Peninsula.

With a trendy Latin style, Yucca's stylish interior design is characterized by the rich, overwhelming color scheme. The wall is vividly painted blue and pink, while the floor is laid with a random pattern of blue and white geometric mosaic. The photo of a lady with blue background extends from the entrance to the bar top. The spiral staircase leads you to the private space on top floor, where you can overlook the whole bar, feeling the passionate atmosphere.

The marble bar with the inscription "DESEO FUERZA AMOR LUJURIA" embodies the spirit of the space. Color combination, pattern on the floor, natural light arrangement and other elements are working together to promote close interaction among the people enjoying their time in this room. Yucca creates a cheerful and exciting place with intimate social atmosphere.

"We imagined a place where one can go to be inspired, a place where your imagination can run free and you can be inspired by your friends," Yucca's charming, modern Mexican feel was born from the thoughts of Salvador Dali, Diego Rivera and Frida Kahlo.

Spanish-style custom-made mosaics, vibrant and eclectic blues, tall candelabra, elaborate flooring, paisley armchairs, iron gates, a number of elements create a warm, cozy and harmonious Yucca, with the spirit of exuberance.

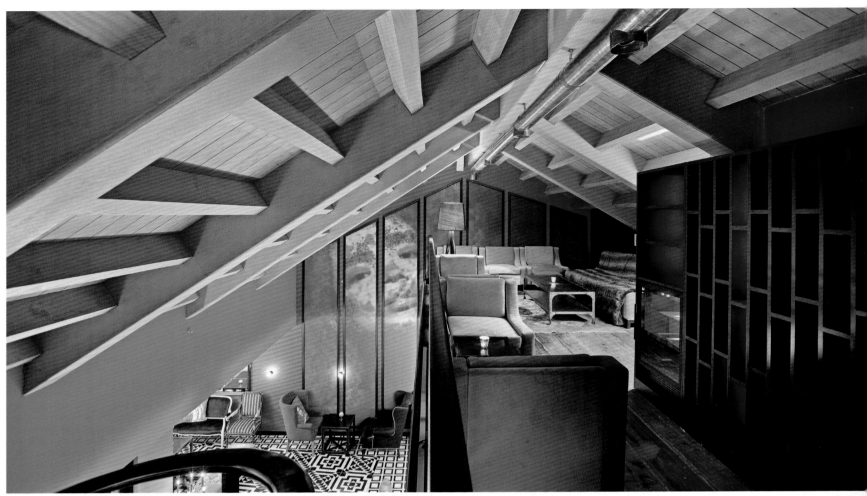

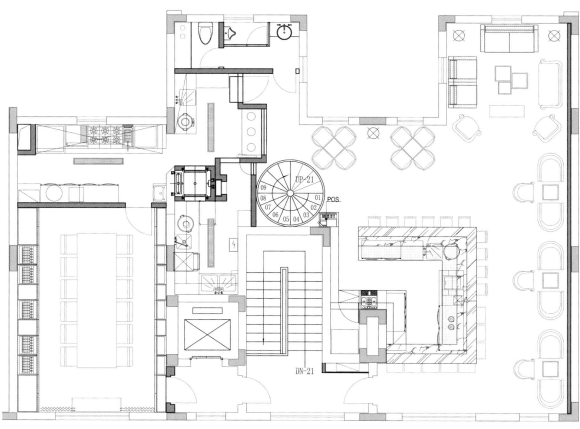

　　丝兰休闲吧坐落在思南公馆26层，在上海旧时法租界的一幢醒目的传统风格的建筑中。这幢楼是丝兰休闲吧的创始人，著名的希腊裔澳大利亚厨师大卫·拉里斯的总部，也涵盖了他的另外三家餐厅：欧立威希腊休闲小馆；12 Chairs，新的高级餐厅及 The Funky Chicken，一家时髦快餐厅。坐落在三楼，丝兰休闲吧是一个时髦的以现代摩登氛围和抓人眼球的设计装修为特色的地方。

　　"墨西哥"这个词和酒吧联系起来总会让人想到一系列固定的套路：仙人掌、宽檐帽、插袋手枪、子弹带、暴露出人造砖块的裂开的灰泥墙或者为了所谓的"上档次"而放置的弗里达·卡洛肖像画。两位设计师托马斯·达里埃尔和伯努瓦·阿福耶尔并没有采用任何的固定套路的元素。区别于一般的老套，丝兰休闲吧的设计向拉丁美洲和伊比利亚半岛丰富的视觉文化致敬。

　　伴随着流行的拉丁式风格，丝兰休闲吧时髦华丽的室内设计以富有冲击力的色彩搭配为特色。墙面用生动的蓝色和粉红色粉刷，地面为随机铺设的蓝白相间的几何图案的马赛克。一幅蓝色背景的女性照片从入口一直延伸到酒吧顶层。旋转楼梯引导客人走向顶楼私密空间，在这里可以俯瞰整个酒吧，感受热烈的气氛。

　　刻有警句"DESEO FUERZA AMOR LUJURIA"的大理石吧台体现了空间的精神。色彩的搭配、地板的图案与自然光的使用等元素使客人在这个空间里能有更多的互动。打造亲密的气氛，丝兰休闲吧创造了令人愉悦兴奋的社交氛围。

　　"我们希望打造一个能激发灵感的场所，在这里，想象力可以自由驰骋，朋友们也可以互相寻找灵感。"丝兰休闲吧迷人、现代的墨西哥感觉是受到了萨尔瓦多·达利、迭戈·里维拉和弗里达·卡洛的启发。

　　西班牙风格的定制马赛克、鲜明不拘一格的蓝色枝状大烛台、精制地板、螺旋纹手扶椅及铁门等一系列元素打造了一个充满热情又温暖、舒适、和谐的丝兰休闲吧。

Zebar

Designer: Francesco Gatti
Design Company: 3 Gatti
Location: Shanghai, China
Area: 569 m²
Photographer: Daniele Mattioli

This project was born in 2006 when a Singaporean movie director and an exmusician from south of China decided to open a live bar in Shanghai. The budget was very low but the client was incredibly good and open-minded to us.

The schedule was very tight and fortunately they liked immediately one of the first concepts the designer proposed to them: a caved space formed from a digital Boolean subtraction of hundreds of slices from an amorphic blob.

The idea looks complex but actually very simple and born naturally from the digital 3D modelling environments where the designer and others enjoy playing with virtual volumes and spaces. The space was subdivided into slices to bring it back from the digital into the real world; to give a real shape to each of the infinite sections of the fluid rhino nurbs surfaces.

In Europe the natural consequence of this kind of design will be giving the digital model to the factory and thanks to the numeric control machines cut easily the huge amount of sections all different from each other.

But in China where the work of machines is replaced by the work of low paid humans. Using a projector they placed all the sections we drew on the plasterboards and then cut each of them by hands. The cost was surprisingly low.

The construction was incredibly fast and was almost finished in a couple of months, when we discovered the naive clients didn't have any business plan and the site remained closed for 3 years and was finally completed and opened in 2010 when they discovered how to run the business.

This is the story of the ZEBAR, a digital design built into an analogic world.

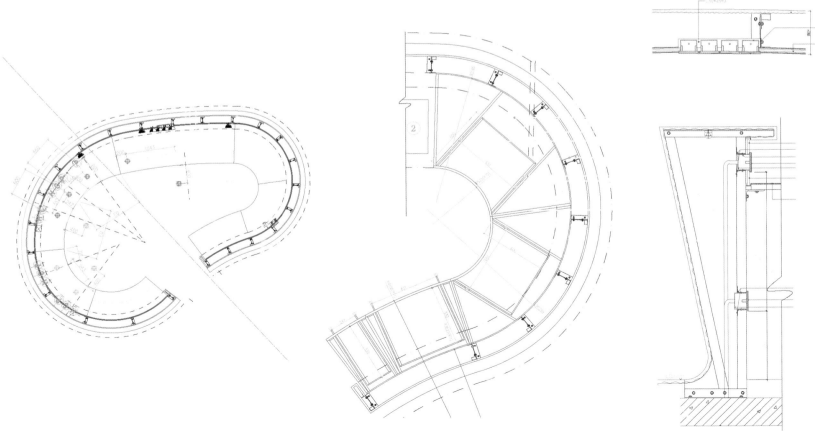

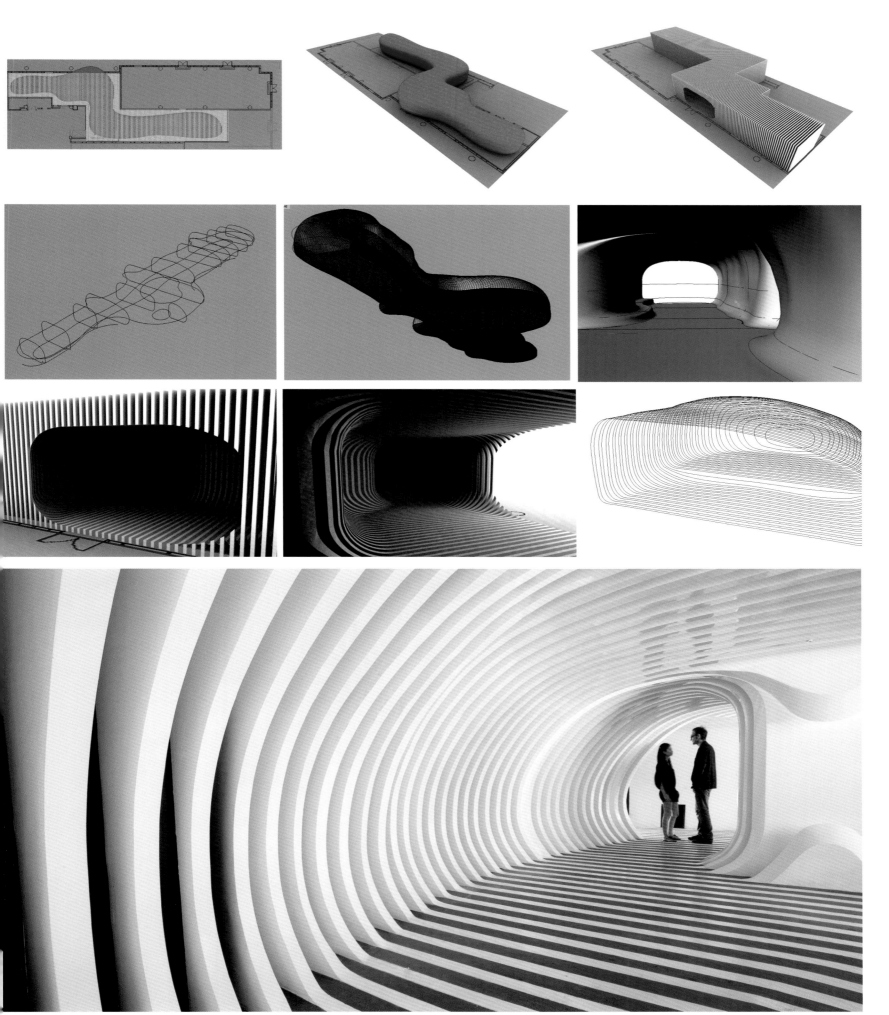

该项目诞生于2006年，当时，一位新加坡电影导演和一位来自中国南方的音乐家决定在上海开一家现场音乐酒吧。他们的预算很少，但思想开通，乐于接受设计师们的建议和想法。

项目设计的时间非常紧迫，幸运的是，客户一开始就设计师提出的几个创意之一表示了喜爱和认同：把一个无形粒子分割成数百片，再对其进行数字化布尔减运算排列，最终形成一个洞穴式的空间。

这个创意表面上复杂，实际上却很简单。当设计师和其他人在数字化三维建模环境里感受操纵虚拟模型和空间的乐趣时，自然而然地就有了这个想法。设计师把空间划分为一片片以实现它从数字世界到现实世界的回归；并为每一个流动的、无限延展的rhino Nurbs曲面赋予一个可视化边界。

在欧洲，这种设计的下一个步骤自然是把数字化模型交给工厂，经由数字化控制，机器可以轻而易举地把许多截面切割成互不相同的形状。然而在中国，机器的工作被人力所取代。利用投影仪，工人把设计师画出的截面投影到石膏板上，然后进行手工切割。这样，设计所需费用低得惊人。

工程的进展出人意料地迅速，大概两个月后完工时，设计师才获悉这两个毫无经验的客户竟还没做出酒吧的商业计划书，就这样，设计又封闭了三年，一直到2010年，当他们终于学会如何经营酒吧时才开业。

模拟世界中的一个数字化设计——这就是斑马现场音乐酒吧的由来。

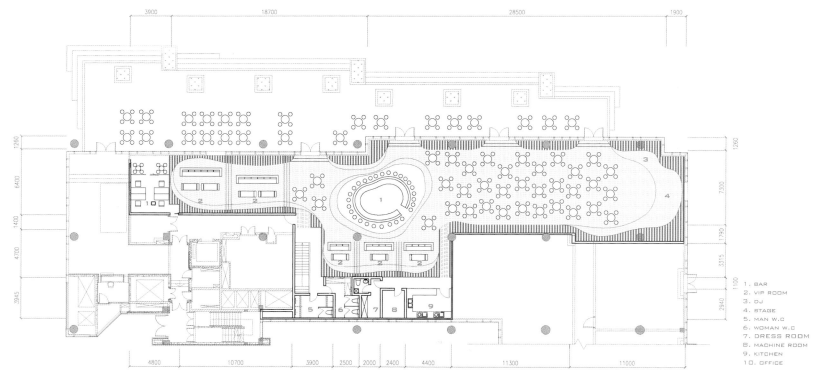

1. BAR
2. VIP ROOM
3. DJ
4. STAGE
5. MAN W.C
6. WOMAN W.C
7. DRESS ROOM
8. MACHINE ROOM
9. KITCHEN
10. OFFICE

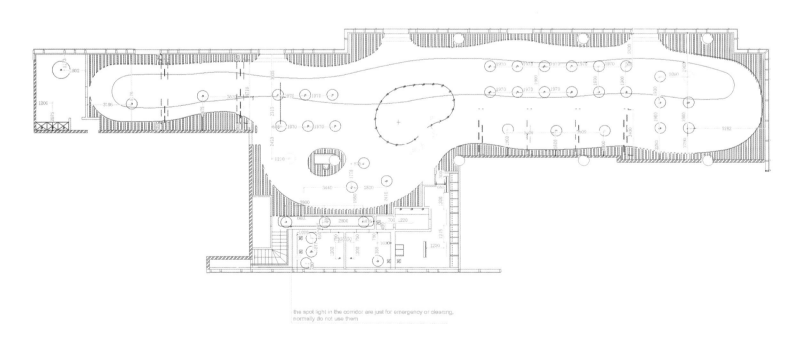

the spot light in the corridor are just for emergency or cleaning, normally do not use them

Plan B

Designers: Abraham Cherem Cherem, Javier Serrano
Design Company: CHEREMSERRANO
Location: Santa Fe, Mexico City, Mexico
Area: 206 m²
Photographer: Jaime Navarro

Plan B is located in Santa Fe neighborhood, one of the most popular zones in Mexico City to go out partying. Plan B is a nightclub-bar of contemporary style that recalls a night out during the 1960s.

The site is composed of a lower level with the bar and lounges, and a mezzanine floor where tables were placed for dining. The services and kitchen are situated at the back of the main floor. The primary interior finishes are black floor and walls made of OSB wood. This type of wood is composed of pieces of different kinds of wood packed tightly together. The main furniture is also designed of this material, giving them a simple but chic look.

In the main floor near the entrance, the bar has different phrases carved into the wall, evoking songs of the Beatles and other iconic rock bands. The phrases change of color according to the LED lighting placed underneath the OSB panels. By placing the LED lighting behind the wood walls, the bar dramatically changes its appearance and colors depending on the time of the afternoon. During the night different tones of light follow the music and DJ set.

In the opposite wall of the bar, a grey curtain hangs from ceiling to floor, delimitating the view from the next door locals. Above the curtain a second layer is designed of cassettes which add texture to the wall. Black paper lamps hanging from the ceiling disguise themselves illuminating the space underneath by giving a sense of floating lights.

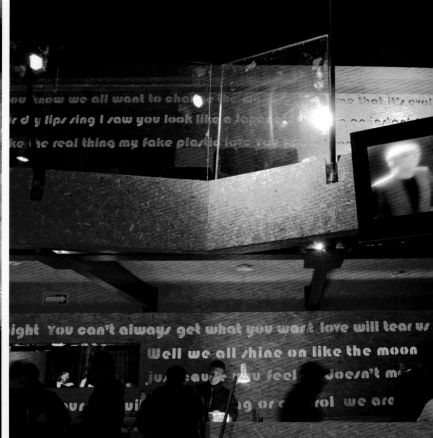

　　B计划酒吧坐落在圣达菲附近,这里是墨西哥城最受欢迎的派对区之一。B计划酒吧是一个充满现代感的夜总会酒吧,却可以唤起人们对20世纪60年代夜生活的回忆。

　　酒吧由低层吧台、休息区和摆放了一些桌子的阁楼所组成。设备区和厨房位于主厅的后面。室内设计主要的外观是黑色地板和欧松板墙面。这种板材由多种不同的木材紧紧挤压在一起而形成。主要的家具也是用这种材料设计制作的,这让它们看起来简单却又别致。

　　在入口处旁的主厅,墙上雕刻着不同的乐句,让人想起"披头士"和其他经典摇滚乐队的歌曲。置于欧松板下面的LED灯使乐句不断地变换颜色。通过在木墙后面安放LED灯,酒吧可在午后随时间的推移而显著改变其外观和颜色。夜晚,灯光还能随着音乐和现场音乐主持人的打碟变换不同的风格。

　　吧台对面的墙上挂着灰色帘子,挡住隔壁客人的视线。帘子的外层利用磁带的设计来增加墙面的质感。悬挂于天花板上的黑色纸灯营造出一种浮灯的感觉,不着痕迹地照亮下边的空间。

Brown Sugar Taipei

Designer: KAN, TAI LAI
Design Company: archinexus
Location: Taiwan, China

Brown Sugar Taipei shop follows the "theater" design concept of its Shanghai shop, which has platforms of various heights in the open space. There are bar area, sofa area, step box area, public dining area, stage area and VIP area. At the front of the base is the elevated bar area, while the public dining area is just at the end of stage area. Both sides are the elevated sofa areas as well as the step seats.

The aisles are widened in the sofa area and furnitures are laid out with a distance to the walls, which gives more freedom and dynamic to the outlines. Stainless steel decorative frames are used for the herringbone pattern screen, the facade and the floor platform. The extensive use of dark tinted mirrors and tawny glass makes the glory of neon light reflected in the space.

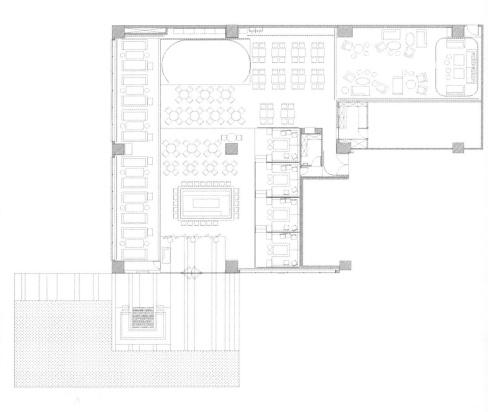

　　红糖酒吧台北店在设计上承袭了上海店"阶梯剧场"的概念，在开放空间内规划多种平台高度，共划分为吧台区、沙发区、阶梯包厢区、散桌区、舞台区和贵宾区，空间前端为架高吧台区与尽头舞台区中包夹着散桌区，左右两侧为架高的沙发区与阶梯式的座位。

　　在加宽的沙发区走道与家具不贴墙的摆设方式下，设计使动线更为自由。人字纹屏风以及立面、地台使用不锈钢材质作为框饰。大量使用的深色镜面及茶色玻璃折射出空间中霓虹光彩的幻影。

Pinon Grill

Design Company: ID & Design International
Location: The Terrace at Town Center, Boca Raton, Florida, USA
Area: 790 m²

A contemporary fusion of Southwestern flavors and modern American cuisine, Pinon is a 790 square meters restaurant and bar designed for an upscale Boca customer yet relaxed in every way for a casual and fine dining experience. The Pinon Trees, synonymous to the Southwestern landscapes are used as an iconic feature fabricated of custom cast bronze sculptures anchoring the four main quadrants of the dining room.

Over-scaled recessed ceiling beams, natural stone walls, and charcoal wood flooring act as the staple room elements. The accent walls comprises of stacked logs with distinct iron artwork of Native American dancers silhouetted against fire "living glass" combined with custom bronzed Pinon Tree sculptures mixed with dramatic lighting and illuminated wine towers capture the imagination and lend themselves to this theatrical approach to experiential dining.

The Pinon Bar/ Lounge features a weathered log community bar, a wall to wall glass wine cellar featuring and a one-of-kind illuminated coral resin bar top.

对西南风味和现代美式菜肴进行融合的矮松吧是一家占地790平方米的餐厅及酒吧，旨在为博卡的高端顾客提供多种方式的休闲美食体验。作为西南风景的代名词，矮松由定制的青铜铸造而成，主要固定在餐厅的四个扇形席间，成为一种标志性特征。

超大型嵌壁式天花板横梁、天然石墙和木炭地板构成了房间的主要元素。主题墙由叠置原木配上美洲原住舞者形象的风格独特的铁制艺术品构成，在火一样的玻璃背景映衬下凸显轮廓。专门定做的铜制矮松雕像，辅以舞台感的灯光和发光的酒塔，引导着人们的想象力，使人产生一种仿佛置身于舞台之中的用餐体验。

矮松吧的特色之处还在于它内设风化原木大众吧、一个配备齐全的玻璃酒窖和一个独一无二的发光珊瑚树脂吧台台面。

Velvet Bar

Designer: Robert Majkut
Design Company: Robert Majkut Design Studio
Location: Multikino Golden Terraces, 59 Złota Street, Warsaw, Poland

Both in the upper part of the Premiere Hall and in the first storey connected to it, there is a VIP zone called the "Velvet Bar". It includes a distinctive long bar and a separate stage – to be used both for music performances and as a separate bay with mobile seats. For a total convenience, the VIP zone has its own cloakroom and toilets, which are also arranged in a luxury and unconventional style. In the monochromatic black and purple interior, walls covered in a quilted plush fabric, mirrors, black stone, and spectacular lighting determine the extraordinary, luxury and stylish climate of the place, which is designed in a special way and takes the reception of film art to the highest global standards. This place aspires to be a meeting venue not just for sophisticated film lovers, but also for the creators and participants of film art.

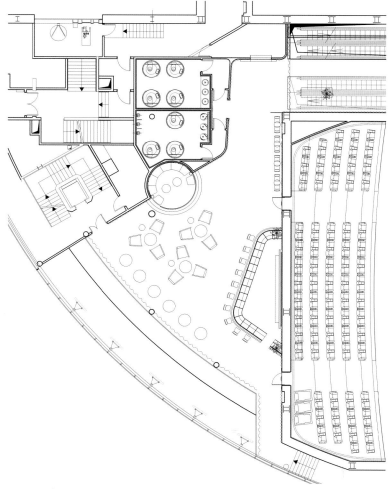

在首映厅的上方和与首映厅相连的第一层，设有一个贵宾区，名为"天鹅绒酒吧"。酒吧里设有一个造型独特的长吧台和一个独立的舞台（可用于音乐表演或作为一个有可移动坐椅的独立区域）。为保证整体上的方便性，贵宾区域有独立的衣帽间和卫生间，这些设施既豪华又不落俗套。在以黑色和紫色为主的内部空间中，墙壁装饰为绒面软包，饰有镜面和黑色页岩。华丽的灯光照明营造出超凡脱俗而又奢华时尚的氛围。本案设计独特，将电影娱乐业的接待规格提升到了国际最高境界，力图打造一个欢聚的场所，无论是高雅的影迷，还是电影业界人士都能陶然于其中。

The Carbon Bar

Design Company: B3 Designers
Location: London, UK

The Carbon Bar was the new addition to the Cumberland Hotel that was completed in April 2007, which also houses Gary Rhodes' W1 restaurant, designed by Kelly Hoppen. The bar is based along Old Quebec street, just off Oxford Street. The interiors inspiration comes from the Shoreditch style industrial architecture. B3 Designers was approached by The Guoman Group to design a new bar.

The general materials used to create this industrial interior were a combination of concrete, brick, steel, mesh and leather. This interior was designed to increase the amount of space available for the clients and have private areas but at the same time have everything visible.

The bar has been designed so that it has descriptive words that would be found in industrial sites. The bar itself is 14 metres long, providing a vast amount of space for people to be able to get drinks from, along with a mezzanine floor, suspended above the lower floor which has a champagne bar. When going upstairs, there is a featured wall of champagne, which is two-storey high, which is filled with some of Taittinger's most expensive and rare bottles. Another feature wall has barrels along it adding to the industrial look.

There is a chain room that is on a raised floor, surrounded by huge steel chains hanging from the ceiling to the floor which hosts the VIP area, within that area is a bar so that the VIPs can stay within that area.

Sean Pines took the photographs of the Carbon Bar, capturing the atmosphere of the bar and the effects of the lighting that illuminate the features within the interior.

The interior design of The Carbon Bar is very industrialised, bringing the East to the West End of London.

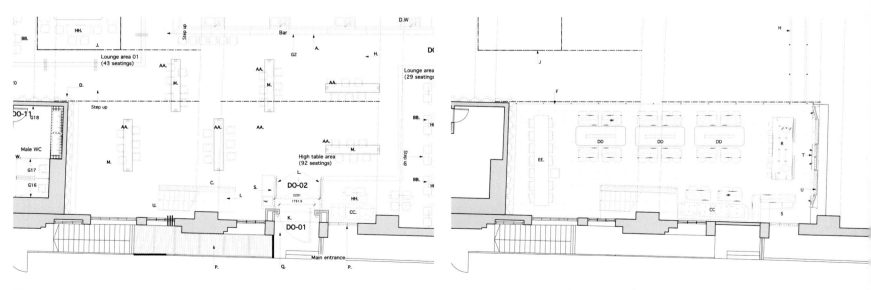

　　碳元素酒吧为坎伯兰酒店新增服务项目，于2007年4月完工。由凯利·赫本设计的盖瑞·罗德的w1餐馆也属于这家酒店。位于老魁北克街上的这家酒店紧邻牛津街。酒吧内部设计的灵感来自于东伦敦风格的工业建筑，由国丰集团委托B3 Designers设计。

　　具有工业风格的内部空间综合运用了诸如混凝土、砖块、钢结构、筛网和皮革制品等普通材质。这样的设计为顾客增加了空间的利用率，既有私密区域又整体通透。

　　酒吧的设计可以用工业厂房的语言描述：全长14米、为顾客提供宽阔的空间。悬于酒吧下层上方的夹层空间是一个香槟吧。步上楼梯，就会看到一面2层的香槟酒陈列墙，墙上满是泰坦瑞香槟最名贵稀有的品种。另一面墙为酒桶陈列，同样增加了整体的工业风格。

　　酒吧有一个钢链围成的贵宾区。该区的地板略为抬升，周围装饰有自天花板悬垂而下、长及地板的钢链。在贵宾区有独立吧间，客人可在贵宾区享受服务。

　　肖恩·派恩斯为碳元素酒吧拍摄的照片，突出了酒吧气氛和其内部空间的独特照明效果。

　　碳元素酒吧的内部设计极具工业化特征，给伦敦西区带来了东区之风。

ircus

Design Company: gt2P
Location: Enjoy Santiago, Chile
Materials: Thermoformed ABS 2 mm, white and chromed, mdf 9 mm plates
Photographer: Aryeh Kornfeld

This project is based on a collection of tiles that create the wall cladding of the bar. It is inspired on the modification of a surface by the action of a force. Using the idea of blowing and its irregular power create a wave as it intersects with the flat surface of the wall.

There are tiles of variable height organized on a grid, displaced as a result of the same power and distributed depending on the distance between the wall and the surface of the blowing. In order to accentuate the movement, each mosaic has cracks that reinforce its curvature and fit in with the others. The mold was made in mdf and machined with five processes to optimize the thermoforming action and maximize the effect of the lines.

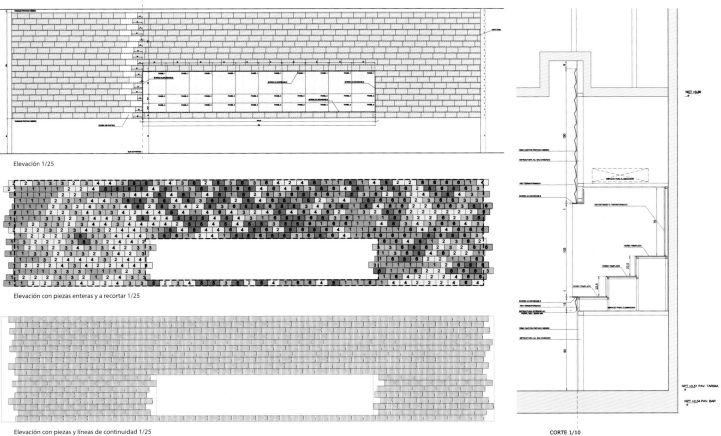

Elevación 1/25

Elevación con piezas enteras y a recortar 1/25

Elevación con piezas y líneas de continuidad 1/25

CORTE 1/10

　　本案的基本理念是要打造一间由一系列瓷砖覆墙的酒吧，其设计灵感来自于以外力改变某种表面的效果。以吹起的方式造成不规则受力所形成的波浪效果同平整的墙面形成立体的意象。

　　瓷砖高低不一，排于网格之上，由同方向受力而产生位移的效果。这些瓷砖的分布方式决定于墙面与吹起的表面的距离。为突出动感，每片瓷砖都做成了纹裂的效果以强化曲面感并同其他瓷砖浑然一体。装饰模型以中密度纤维板机器制作，历经五道工序，使热成型与线条的效果都达到了极致。

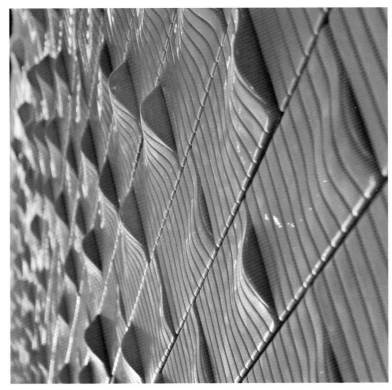

Meltino Bar & Lounge

Designer: Cláudia Costa
Design Company: LOFF
Area: 180 m²

Meltino Bar & Lounge is a place intended to update the concept of a coffee embedded in a shopping center. Having the strong presence of the design element, the furniture is the result of coffee derivatives, in order to allow that the community assimilates the singularity of a place that serves as a lounge and bar.

Both are similar and transparent, although with different purposes. The lounge site has a relaxed atmosphere and offers gourmet products. The bar site is appropriated to drink coffee.

Cláudia Costa, LOFF atelier mentor, conceived this project from the geometrization of coffee grains, always with the concern to exclude the idea presented in the space to the final consumer, that they are drinking coffee in a commercial gallery. The final result is the invasion made by the coffee grains instead of the Shopping layout.

Meltino Bar & Lounge is a coffee grain that peeks, pierces walls, roofs and counters.

In the first meeting with the client, it was asked to draw a charismatic and unique project for a bar with the identity of the brand Meltino.

The idea was to create a project that endured in the visual memories of the public. The first intention was to build a bar inside a mall without a sense of enclosure in a confined space. The thought of how to be inside and have the feeling of being outside was always present. The box inside the box, the light, the transparence, and the defragmentation of the space, all the elements are associated with the simple act of drinking coffee.

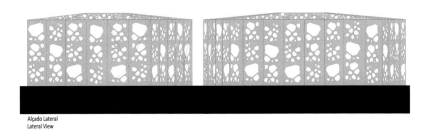
Alçado Lateral
Lateral View

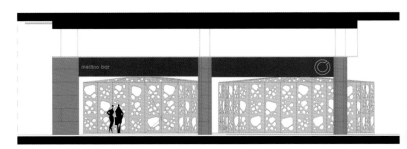

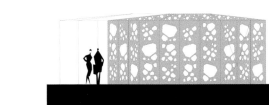
Alçado Frontal
Front View

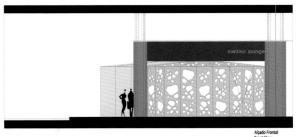
Alçado Frontal
Front View

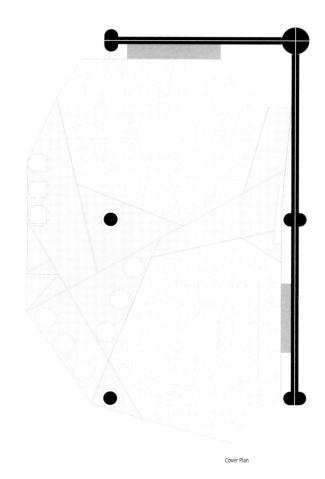

Cover Plan

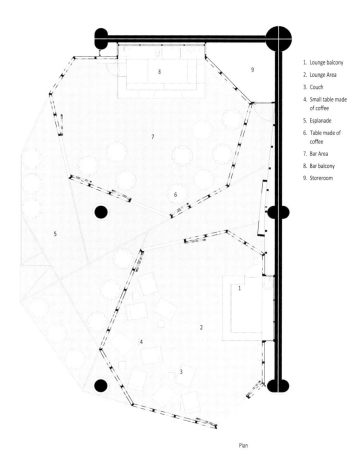

1. Lounge balcony
2. Lounge Area
3. Couch
4. Small table made of coffee
5. Esplanade
6. Table made of coffee
7. Bar Area
8. Bar balcony
9. Storeroom

Plan

　　玛提诺咖啡吧与休闲吧重新定义了在购物中心里享受咖啡的感觉。本案家具都与咖啡相关，设计元素具有很强的存在感，使得咖啡吧和休闲吧的独特性能够很好地融为一体。

　　本案咖啡吧和休闲吧的设计相似且透明，尽管二者功能不同。休闲吧气氛闲适，且提供美食；咖啡吧则更适合品尝咖啡。

　　LOFF工作室的设计师克劳迪亚·科斯塔构思本案的灵感来源于咖啡豆的几何形状，力争让消费者身处此空间时感受不到自己是坐在商业空间里喝咖啡。本案让消费者最终感受到的不是商场的布局，而是四周的咖啡豆。玛提诺咖啡吧与休闲吧就像是一粒咖啡豆，窥视着、穿透了周围的墙壁、屋顶和柜台。

　　客户与设计师第一次会面时，提出要设计一间魅力非凡、风格独特且拥有玛提诺品牌特色的咖啡吧。

　　设计师的想法是创建一个能在人们的视觉记忆中停留的设计。首要目的是在商场里建立一个无空间密闭感的咖啡吧。人们常常思考怎样使置身室内的人有一种身处室外的感觉。只要在这里品尝着咖啡，就仿佛是在空间内的另外一个空间里，感受到光、透明以及时光的交错。

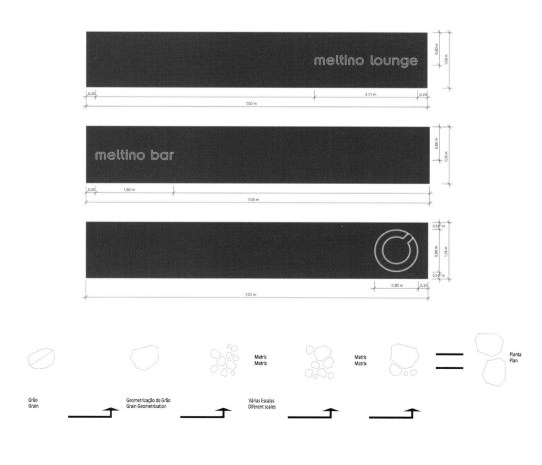

| Grão | Geometrização do Grão | Várias Escalas | | Planta |
| Grain | Grain Geometrization | Diferent scales | Matrix / Matrix | Plan |

Oxo Tower Bar

Designer: Shaun Clarkson
Design Company: Shaun Clarkson ID

The designers wanted to see the redesign as a whole, introducing a series of fluid shapes that swam around the building, softening corners and straight lines. The bar was originally designed with a central core housing the kitchen, toilets and service areas situated in a glass-roofed structure. They wanted to contemporise the design, with boat shaped fluid serviceable counters that housed state of the art services all relating to each other.

The new bars, receptions and coat checks are sculptural forms that were shaped in all directions. The designers wanted to use light, bright hues, so the choice of white Corian seemed obvious. Being a costly material, it was a challenge to create the illusion of solid blocks of white sculpted Corian without the cost and weight involved in using a solid material. To achieve this, a series of formers were created, that in turn were clad in formed sheets of Corian. These were built in modular units in a workshop and assembled on site. Then, the seams were filled and polished.

The internals of the bars were fitted with formed and modular stainless steel units that had to make this busy bar super efficient.

This needed to present the extensive range of premium spirits that are needed to mix the cocktails that Oxo are renowned for. To reflect abstractly the light, movement and vigor of the lovely Thames, the designers used a reflective stainless steel tile that was mirror clear, but small enough to abstract and deflect the light, to reflect the mood of London's diverse pallet of light. This was twined with glass and steel to give a feeling of infinity.

此次改建设计师们试图进行一体化设计，采用一系列流线形状，使其环绕建筑本身。酒吧原本设计有一个位于中央的核心区，内有厨房、卫生间和服务区并带有玻璃天花。设计师们的初衷是将该设计与具流线动感的船形服务台融为一体，而后者集合了彼此关联的各种高品质服务。

新增设的吧台、前台和衣帽寄存处采用雕塑的形式，面向四面八方。设计意图是利用淡雅而明亮的色调，因此，白色可丽耐材料似乎就成了最佳选择。此种材料造价较高，可若不使用这种造价高、重量大的材料而又要产生出实心可丽耐固体雕塑的视觉错觉的确是个挑战。为了达到这个效果，需要先制作一套内模，然后在外面覆盖可丽耐板材。这些都是在工厂将各个模块制作完毕后在酒吧实地安装的，之后还要对接缝进行填充、打磨。

酒吧内部均用不锈钢的组合部件装饰，使得这家繁忙的酒吧达到效率最高的工作模式。

这与制作Oxo享负盛名的鸡尾酒所需的各种昂贵烈酒关系密切。为了能够写意地表现明快的泰晤士河的光影、流动和活力，设计师们使用了折射率较高的不锈钢贴砖，这些贴砖虽小但如同明镜一般展现出伦敦纷繁多变的灯光。装饰与不锈钢材质相映成趣，给人一种空间无限延伸之感。

Press Club Urban Wine Venue

Design Company: BCV Architects
Location: San Francisco, California, USA
Area: 762 m²

The charge for this innovative wine collective/tasting lounge was to create a sophisticated urban identity while simultaneously realizing a modern expression of the spirit of California wine country.

Within a clean-lined, minimalist aesthetic, the architects have captured the casual sophistication of the wine country and created a space that is both timely and timeless. Taking cues from Napa's unique blend of industry and natural beauty, the restrained design celebrates the juxtaposition of the industrial against the organic as a foundation for the projects expression.

A sense of place has been created within an otherwise challenging urban location. The majority of the project's 762 square meters is housed within the subterranean cellar of San Francisco's Four Seasons Hotel. The cellar program includes individually-maintained tasting bars for 8 regional world-class wineries, public and semi-private lounge spaces, a central wine and food bar, a private event room and support spaces.

A bold monolithic concrete stair dramatically links the upper level arrival and wine shop to the extensive wine-centric area below, flanked by a 5.18 meters tall LED back-illuminated bottle wall. This signature feature wall transcends the divide between entry level and cellar and punctuates the project with its striking color patterning.

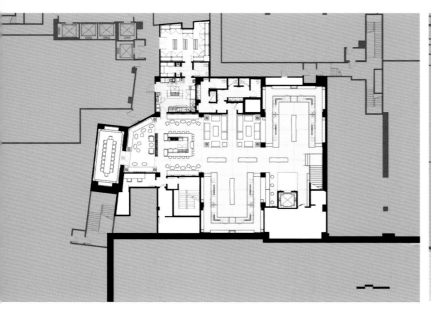

145

本案集收藏和品酒于一体，其创新的设计理念旨在体现出高雅的都市品位，并对加州酒城之精神赋予现代的诠释与传承。

清晰的线条结合着简约的美学元素，建筑师在每个不经意间展示着酒城独特的优雅，也创造出一个既具时代感又亘久长存的空间。纳帕的工业与自然美景奇妙融合，本案由此获得设计灵感，控制得当的设计彰显出工业与自然的和谐共存，这也是本案艺术表达的基础。

这里虽是充满挑战的市区，但酒吧给顾客以归属感。本案占地762平方米，项目主体位于旧金山四季酒店的地下酒窖内。酒窖工程包括为8个地区的世界级酒庄单独运营的品酒酒吧，公共和半私人的休憩空间，一个提供葡萄酒和食品的中央吧台，一个私人活动室和许多辅助空间。

从顶部入口沿着一个设计大胆的混凝土楼梯而下，进入酒窖，就会来到豁然开朗的葡萄酒中心区，侧面设有一个5.18米高的LED背照式酒瓶墙壁。这面特色酒瓶墙壁实现了入口与酒窖间的完美过渡，其醒目的彩色图案也使得酒窖熠熠生辉。

Na Mata

Architects: Juliana Nohara, Flavio Faggion
Authors: Fernando Forte, Lourenço Gimenes and Rodrigo Marcondes Ferraz
Coordinators: Ana Paula Barbosa, Sonia Gouveia
Trainees: Bruno Milan, Carolina Matsumoto
Lighting Consultant: Castilha Iluminação
Engineering: Foz Engenharia
Location: São Paulo, Brazil
Area: 900 m²
Photographer: Fran Parente

Before becoming was the bar so called Popular, and prior to this, in the 1970s, it was the well-known Clyde's bar.

The designers' proposal had to be based on the maintenance of certain pre-existing aspects due to matters of execution deadline – as a consequence, the kitchen and some other items have been kept back to their originals. They opted to make a more radical change in the rest using materials, finishings and illumination in order to create a clear identity for the venue and its different areas.

The restaurant was completely remodeled – the entrance access, all made of glass, allows one to see through all the movement inside. The facade, made of cement finishing, is more discrete. The long bar has been kept, but in the foreground there is a long panel made of "butcher's little letters" writing the menus, the bands of the week, curiosities and other activities from Na Mata. This element have been reinterpreted and applied in a new, much more emphasized way.

The walls and the ceilings have received stickers with a standard of orange circles, graphic design developed to be the new identity of Na Mata. This graphic design has also been applied in the concert area and on the second floor, unifying the venue, going in an opposed direction of that of the former proposal. The circles are orange with varied diameters in order to assume a random standard on the vinyl sticker and giving a look to the whole edifice internally. The illumination is integrated to this standard, with luminaries always occupying the space of some circles.

A reorganization of the back, the concert area of the nightclub was very advantageous: the stage have been expanded 50% and the restructured dance floor remained the same size, but gained in perspective, besides the VIP areas on both sidelines, the most wanted area of the floor currently. A discrete support bar avoids the flow to the front of Na Mata where the main bar is placed. Besides a new, much clearer identity in its different areas, the intervention proposed to Na Mata was able to intensify its previous vocations – much more organized concert venue, ready to receive all sorts of musicians and bands.

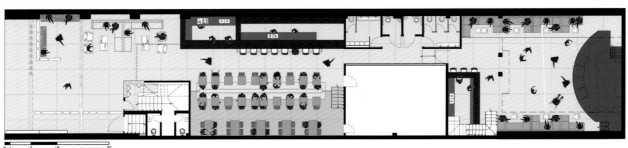
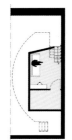

SUBSOLO
ESCALA 1/100

在成为娜玛塔咖啡吧之前，这里是一间名为"流行"的酒吧；再往前追溯，它在20世纪70年代曾是大名鼎鼎的克莱德酒吧。

受施工交付期限制，设计团队的方案不得不以保留某些现有元素并以其为基础——结果就是，厨房和其他一些东西被原封不动的保留了。他们选择在其他的部分进行大刀阔斧的改造，采用材料、涂漆和照明手段将其打造成了一个有鲜明特点的地方。

餐厅被彻底改造——入口全部由玻璃打造，使人们能看清里面的所有的活动。立面涂有水泥涂层，颇具卓然之气。长吧台被保留了下来，但是在其上方有一个长达20米的电子屏幕，写着菜单和本周乐队表演信息以及其他精彩活动等。旧的吧台元素被重新诠释，并以一种全新而引人注目的方式应用了起来。

墙和天花板上贴有橙色的圆形装饰，平面设计已经成为娜玛塔的新标志。在乐队表演区和二楼也采用了这种平面设计，将空间风格统一起来，展现出与酒吧原设计截然不同的风貌。那些橙色圆形装饰直径大小不一，以期在乙烯贴纸上呈现随意之感，为空间内部打造出一种特殊氛围。由于照明物总是安装在圆形图案的位置，空间照明与装饰排列契合统一。

位于内侧的乐队表演区的重新布局非常有效果：舞台区扩大了50%，重装的舞池原有面积保持不变，但在视觉上扩大了。此外，两侧还有贵宾区，那里已成了目前舞池最受欢迎的地方。独立的辅酒吧避免大量客人涌向位于娜玛塔大门口的主酒吧。在娜玛塔作为咖啡吧有明晰定位的同时，这个隔断区也表明，娜玛塔可以延续它之前的功能——一个随时供各类型音乐人和乐队使用的表演场地。

Hyde Lounge

Designer: Gulla Jonsdottir
Design Company: G + Design
Team: Ann Vering
Contractor: DLD Contractors
Location: 8029 Sunset Boulevard, Los Angeles, USA
Area: 139 m²
Photographer: Ryan Forbes

Originally opened in 2006, Hyde Lounge was instantly dubbed "the most exclusive club in America," by the Los Angeles Times. Following an extensive re-imagination of the legendry space, Hyde Lounge's new design aesthetic is reminiscent of an ornate jewelry box, holding the treasures of Hollywood in its protective grasp. Inside, champagne leather decorating on the walls and ceiling with lights behind create an alluring movement of shadows. Adding to the intrigue is a veil of metal rebar that forms a passageway and wraps around the columns of the interior. Crowning the bar is an installation of cut jagged wood cubes encased by white curved plaster and black mirror that encloses the space and invites guests to engage.

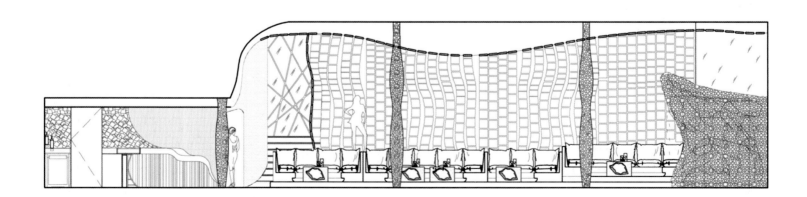

INTERIOR ELEVATION 1
SCALE: 1/8"=1'-0"

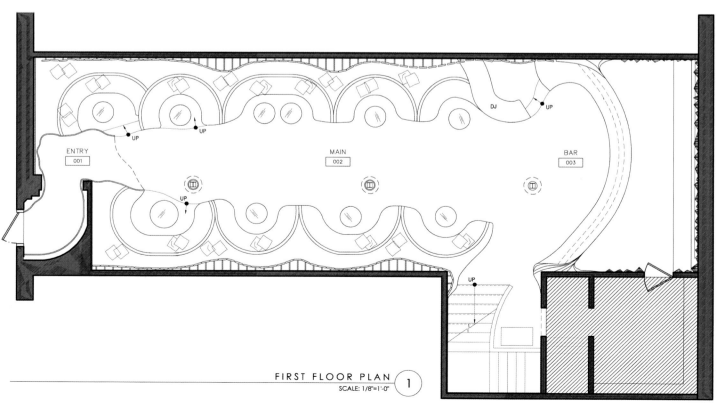

FIRST FLOOR PLAN 1
SCALE: 1/8"=1'-0"

2006年最初开业之际，海德休闲吧即被洛杉矶时报评为"美国最独特的会所"。在对这个传奇空间进行大量的重新构想之后，海德休闲吧崭新的设计美学让人联想到华丽的首饰盒，里面装有好莱坞的宝藏。在休闲吧内部，香槟色皮革装饰于墙壁与天花板上，其后面有灯光装饰，制造出一种光影晃动的迷人效果。除此之外，穿过一个金属螺纹钢构成的通道，客人可以直达休闲吧，而且内部柱体也是由这种材质装饰而成。切割参差不齐的木质方块表面饰有白色弧形石膏以及黑色镜面，将空间封闭起来，吸引着客人驻足于此。

Cienna Ultralounge

Designer: Antonio Di Oronzo
Design Company: bluarch architecture + interiors + lighting
Location: Queens, New York, USA

Cienna Ultralounge is designed to offer the ultimate entertainment experience to the discerning New York crowd.

Cienna Ultralounge is an essay on softness. This project is conceptually linked to the mathematical clarification of the shape of a silk cocoon via a cosine Fourier series. Formally, the space transposes such geometric formalization into a meandering system of openings on the ceiling and walls.

The space is a cocoon entirely made of a seamless, tufted upholstered shell. The distribution of the 8,888 buttons defining the tufting was organized via 3D software, and allowed for precise fabrication and permissible tolerances. The system of openings in the ceiling and walls of this shell is punctuated by 88,888 acrylic strands. Behind them is a system of LED light fixtures softly flooding the strands and the space with warm hues, responding to music impulses via ad-hoc software.

The space has reachable boundaries that offer a soft, tactile, experience. The seating is made of sensuously outlined booths continuing the tufted shell. The tables are custom-made in a sumptuous, full profile of poplar wood, and are lacquered in a gloss finish. The 13.72 meters bar shares its design with the bar at Cienna Restaurant. It is a lighted prism of honey onyx, and balances the DJ booth across the venue. The DJ booth also functions as a stage for live performances, as its sides are removable. The sound system is state-of-the-art and is fully integrated in the tufted shell. The sound system, the lighting system, and all other design systems are experientially and technologically interconnected, as they exist and support one another to offer a seamless overall performance.

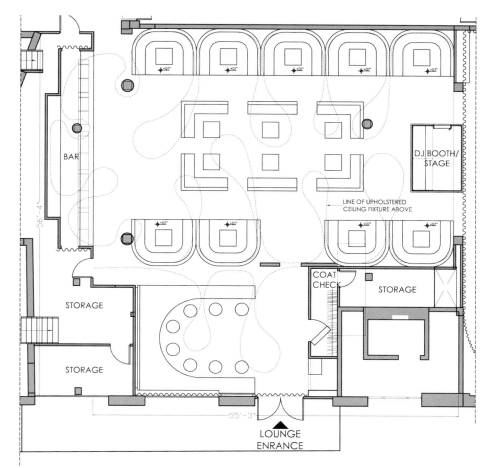

森娜高级休闲吧的设计旨在为有品位的纽约大众提供一种终极的娱乐体验。

森娜高级休闲吧像是一篇情感柔和的散文。本案设计理念通过余弦函数和蚕茧的数字模型联系起来。空间设计将这种几何外观转化成一个蜿蜒的开口系统，开口主要设计在天花和墙壁上。

这个空间就像是一个蚕茧，完全被一个无缝的穗饰外壳所包围。界定穗饰外壳的8888个按钮通过三维软件设计布局，构造精确，仅存在一定允许偏差。天花板和墙壁的开口系统由88888根塑胶绳支撑。在它后部安装的是LED灯系统，为整个室内空间带来温暖的光线，与点对点软件系统调控的音乐相互呼应。

空间的可达边界给人一种柔和、具有触感的感觉。坐椅是感观上界线分明的独立隔间，与整体空间融为一体。桌子是定做的，主要材料是杨木，再加上光滑的抛面，凸显出豪华。13.72米的吧台同森娜餐厅的吧台设计相似，它是一个松香黄的发光棱柱体，和对面的音乐主持人台相对。因为边侧可移动，音乐主持人台也可以作为现场表演的舞台使用。音响系统是最先进的，充分融入了穗饰外壳之中。音响系统、灯光系统与其他设计体系在经验和技术上都是内部相通的，因为它们共同存在，相互支持，呈现出的是一个完全统一无缝的空间。

Dusk

Designers: Lionel Ohayon, Siobhan Barry
Design Company: ICRAVE
Location: Atlantic City, New Jersey, USA
Area: 929 m²
Photographers: Frank Oudeman, Alan Barry

Dusk at Caesars Atlantic City is layered with the best technology and technology inspired materials. The design of the space was inspired by the descent down into the space from the floor above, and by its proximity to the Ocean. From the moment you step through the perforated metal entrance and descend down the stairs, you feel like you are being transported to a different world — a dark adult underworld playground. The club is arranged in concentric ovals, like the rings of a water droplet around the main dance floor. You enter on the outer ring of the club that is elevated 0.30 meters higher than the dance floor allowing guests to view the scene. From here you can walk around the outer ring to the bars, to the middle ring table service area, or dive right onto the dance floor.

The color palette emulates the colors of dusk with different shades of blue and copper. The space is layered with various textures supplied by perforated and polished metals in some areas and the rest of the areas offering a warm and luxurious feels with leather wrapped handrails, plush fabrics, and solid woods. The cool blue and purple hues taken from the night sky are offset with the warm amber light given off by the backlit onyx and bronze mirror throughout the space.

The main design feature and focal point in the space is a multilayered undulating mesh ceiling that hangs over the entire heart of the club and measures 232.26 square meters Impressive theatrical lighting is embedded in the various layers of the mesh ceiling and is placed on lifts that allow the form to be illuminated from inside or to be projected on from the outside. The ceiling feels as though it is actually moving. Curving up and over the elevated DJ booth and stage are nearly a thousand color-changing lights giving an amazing backdrop for the DJ. The lighting in the entire space can be controlled from the DJ booth — the whole room can correspond to the music with the push of a button. The custom designed audio system is unrivaled in acoustic output and fidelity.

As a complete departure and contrast to the experiential design of Dusk, Dawn, the VIP area is a modern part in a small old-world pub that sits below the main club. Dawn is a small intimate space wrapped in warm burgundy leathers with aged copper sconces.

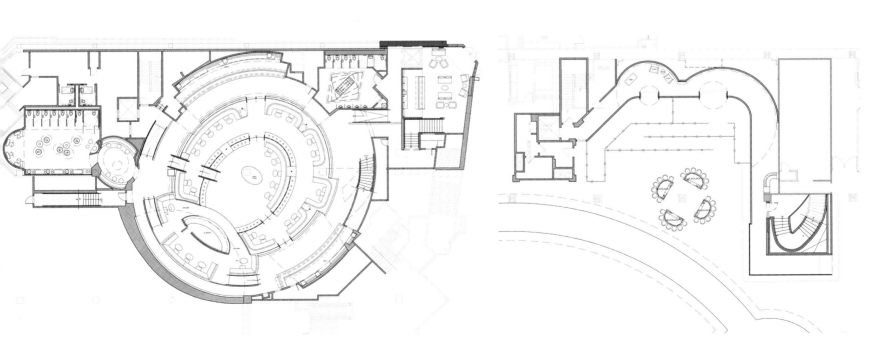

黄昏酒吧位于大西洋城凯撒酒店，由顶尖技术和富有科技灵感的材料打造而成。空间设计充分反映了从上层地面一直延伸到内部的垂直空间及其临海的特征。踏入打孔金属板装饰的大门，随楼梯拾级而下，你会感觉如入异境——一个幽暗的成人地下游乐场。本案整体布局呈同心椭圆形，如同层层波纹环绕舞池。最外层环台高出舞池0.30米，可以看到整个场地。沿着外层环台漫步，还能走到吧台和中层环台餐桌服务区，也可以直接移步舞池。

色彩上用浓淡相宜的蓝色和红铜色模拟出黄昏的颜色。整个空间中，一些地方用打孔金属板和抛光金属板装饰出不同的纹理和质感，而另一些地方则用皮革包裹的扶栏、长毛绒和实木进行装饰，营造出一种热情、豪华的感觉。空间中到处都装饰着由背光照亮的玛瑙和铜镜，发出温暖的琥珀色灯光，冲淡了来自于夜空的蓝紫色冷色调。

本设计的主要特色和焦点是一个悬挂于核心地带上方的多层波浪形网格吊顶，总面积达232.26平方米。炫目的剧场照明灯内置于网格吊顶的各层中，也安装在提升装置上，使得吊顶可从内部或外部被照亮。吊顶给人造成一种一直在移动的感觉。位于高处的调音台和舞台的上方笼罩着差不多一千支不断变换颜色的灯，为音乐主持人构建了一个令人叹为观止的背幕。整个空间的照明全部都可以从调音台进行控制，可以根据音乐的节奏，用按钮控制房间的灯光，造成旋转、追逐等效果。音响系统完全是定制的，具有无与伦比的音响效果和保真度。

作为黄昏酒吧实验式设计上的彻底背离和反差，位于下层的贵宾区"黎明吧"则完全是一个具有现代气息的小型怀旧酒吧，一个由温暖的酒红色皮革和仿古铜烛台包裹的小型私密空间。

Tamarai Lounge & Restaurant

Designer: Shahira H. Fahmy
Location: Nile City Towers, Cairo, Egypt
Area: 1100 m²
Photographer: PSLAB

Egypt's history, culture and heritage are re-imagined in this most cutting edge of settings; a luxurious and lavish project in the heart of Cairo. The main concept was to create a space where the magic and mystery of the ancients in Egypt are made modern.

Each design element was especially commissioned and exclusively crafted. The temples of the pharaohs are reborn in soaring pillars molded from amber-colored granite and marble, while the rough-stone walls are imprinted with drawings of temple plans. The textures and colors of the ancients envelop the space. Lush hand-made linens lined with gold tumble down the floor-to-ceiling windows, covering chairs and lining tables.

The formal and the informal blend seamlessly in 400 square meters of fluid space that include fine dining areas up to 250 guests curving around the lounge area and the spots in-between. The floor rises and falls, with carefully created levels ensuring that guests never loose sight of the Nile flowing beneath.

The pièce de resistance is the pharaonic sun-boat that seemingly floats at the very heart of the Tamarai. Made of shimmering wood, it houses the main bar and forms a fantastical focal point with the iridescent curves of the boat stretching up to the ceiling and out to the reception area, creating a map in the sky to draw you into the space.

And beyond it all, past the panoramic walls of glass, another world awaits, right beneath the stars. 500 square meters of outdoor space evokes silent Nile-side dreams of swaying palm trees. The shades and textures of papyrus dominate the space, with olive trees and mute greens creating a lush landscape.

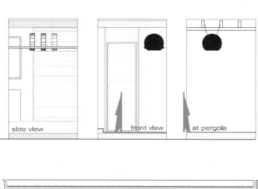
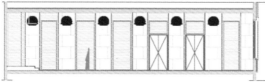
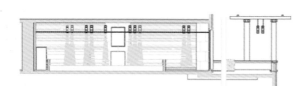
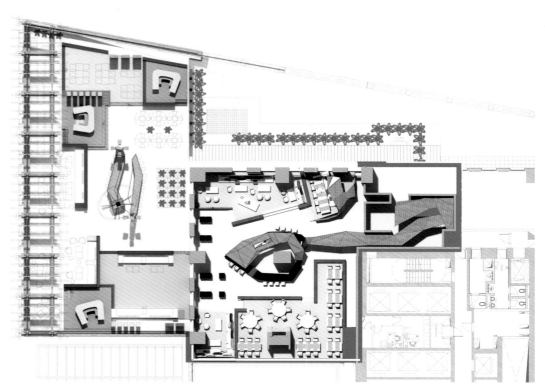
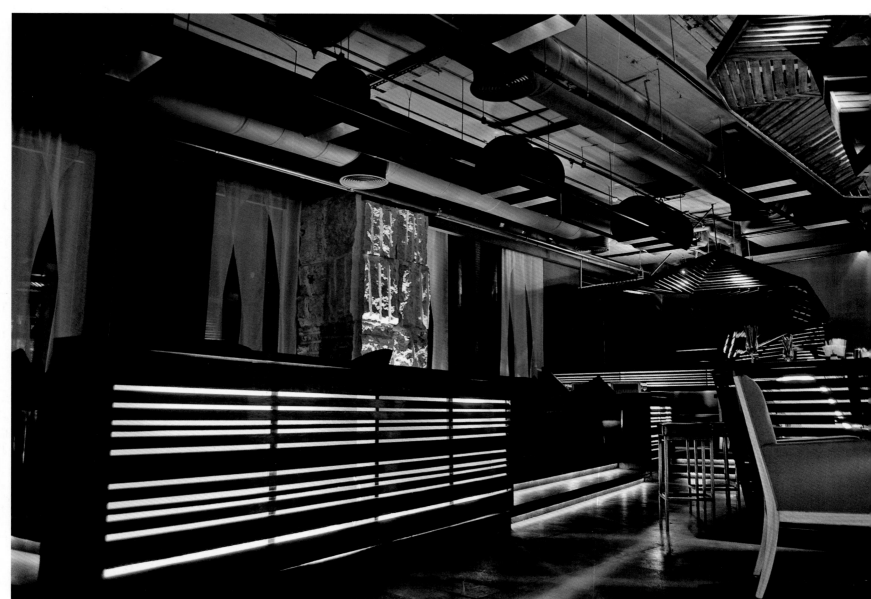

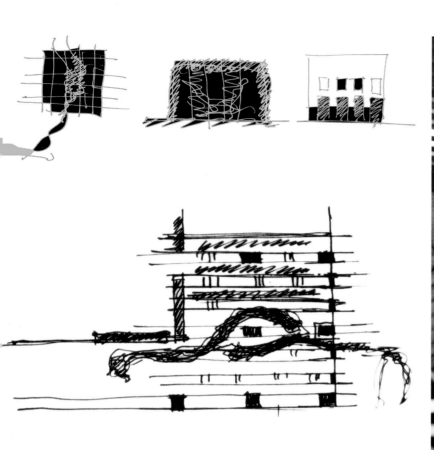

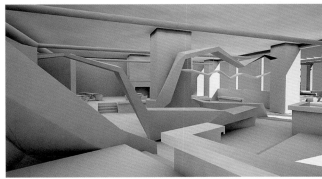

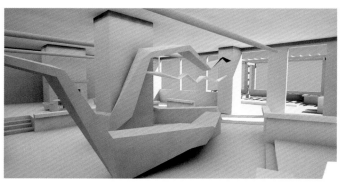

　　本案位于开罗市中心；在这家设计特色尤为鲜明的豪华饭店里，埃及的历史、文化和传统均获得了全新的诠释。其设计理念在于创造一个能为古埃及的魔幻与神秘赋予现代感的空间。

　　设计中的每个细节都是特别定制的。法老庙宇在高耸屹立的花岗岩和大理石石柱中获得重现，原石堆砌成的墙壁上刻画着庙宇的设计图。古埃及风格的花纹和用色是整个空间的基调。精美的金色条纹的手工亚麻布料挂在落地窗上，就连桌椅也是用这种布料装饰的。

　　在这400平方米的流畅空间里，正式与非正式两种风格得到了完美结合。吧台周围分布的就餐区域可供250人就餐。地面高低起伏，被精心设计成恰到好处的高度，可以让客人正好看到下面流淌的尼罗河。

　　餐厅里的法老太阳船引人注目，这太阳船仿佛在项目中心飘荡。发光的木质船体上是餐厅的主吧台也是整个餐厅的焦点。船体发出的光线投射到天花板并延伸到迎宾处，在空中形成了一张地图，指引客人进入餐厅。

　　最特别的是，穿过全景玻璃墙，就在星空之下，这里向你展示另外一种面貌。500平方米的室外空间能诱发客人关于摇曳着棕榈树的静谧尼罗河畔的梦想。纸莎草的颜色和纹理为空间装饰的基调，连同橄榄树和深绿色的装饰，为酒吧营造了一种浓郁的闲适风情。

VYNE

Designers: Rob Wagemans, Janpaul Scholtmeijer, Charlotte van Mill, Erik van Dillen
Design Company: Concrete Architectural Associates
Location: Prinsengracht 411 1016 HM, Amsterdam
Area: 125 m²
Photographers: Ewout Huibers, www.ewout.tv; Concrete Architectural Associates

VYNE consists of two parts. On the right-hand side is the "brown café", with oak floor, walls and ceiling and a long brown leather couch against the wall. This room is reminiscent of the classic French wines, the oak planks serving to give the impression that one is sitting in a wine vat. The left side serves as a wine library, an entire wall of stainless steel elements parts completely filled with wine bottles. The clinical look melds well with new wines from Chili, and Africa; new wines that are often produced using modern technology. Long bar tables adorn the centre, providing a cozy, Burgundian and lively atmosphere and offering places for approximately 100 people. VYNE also offers a peek in the kitchen, which has a "chef's table" where a tasting can be held.

In the oak ceiling the lighting is provided by random placed chrome Bolsters, which makes the lighting dynamic and theatrical.

While VYNE is easily recognizable as a "sister" to Envy, at the same time it has its own character and atmosphere. Even the name VYNE, in the form of an anagram, is a playful reminder of Envy. Where Envy is about food with wine, VYNE will be offering its guests the opportunity to try the ultimate wine-food combinations. The name is also indicative of VYNE's international character, and the fusion of the English words "fine" and "wine" provides an excellent illustration of the concept.

1 covered course
2 entrance
3 check-inn
4 canteen M
5 livingrooms
6 elevators
7 supply
8 artwork
9 screenline inside glasspanel
10 luminaire
11 double glazing with -5, -2, 0, 2, 5 degree inclination
12 black rubber cover fillet
13 black-out curtain
14 gypsum fibre board
15 steelconstruction module
16 zebrano floor/ceiling
17 aluminium window frame, silver/metallic; depth: 250, 350, 450 mm
18 black coated composition
19 aluminium window frame
20 window blind
21 flooring: bamboo parquet/floating floor/ belgian bluestone

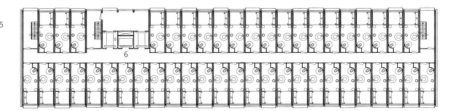

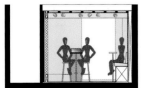
section D

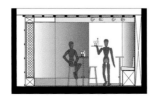
section E

section F

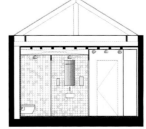
section G

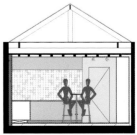
section H

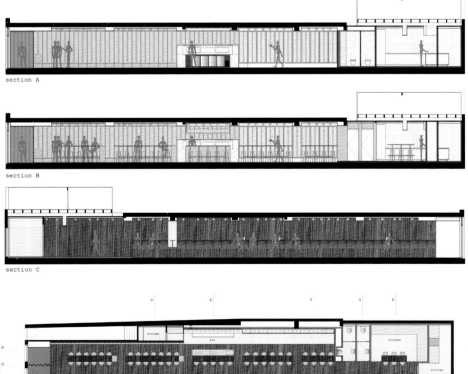

section A

section B

section C

floorplan

维尼由两部分构成。右手边的是"棕"酒吧，店内装饰有橡木地板、墙面和天花板，靠墙摆放着长长的棕色皮革沙发。橡木板使人觉得自己仿佛置身于巨大的红酒桶内，整间房间让人想起古典的法国红酒。左手边是一面红酒展示墙，整面墙上的不锈钢架上摆满了红酒瓶。硬朗外观与来自智利和非洲的新产红酒相得益彰。新产红酒通常都是使用现代工艺酿制的。长长的吧台位于酒吧中央，营造了一种活泼舒适的勃艮第风格，可容纳将近100人。维尼也配备有厨房，在那儿有"主厨餐桌"，可以举行品酒会。

橡木质天花板上随意安置的铬灯具为酒吧提供照明，使得房间的光线显得跳跃而具有剧院风格。

维尼很容易被当成是恩维酒吧的姐妹店，但同时它又具有自己的特点和风格。甚至维尼这名字都是恩维这个字调皮的变体之作。恩维主要是提供食物和红酒，而维尼则是向客人提供红酒与食品的极致搭配。名字本身也暗示了维尼的国际化特色，"卓越"和"红酒"两个词的融合为酒吧的理念做了非常好的诠释。

Numero Bar

Designer: Isay Weinfeld
Project Team: Alexandre Nobre, Juliana Garcia
Collaborator Architects: Domingos Pascali, Marcelo Alvarenga
Project Manager: Monica Cappa
Location: São Paulo, Brazil
Area: 551.32 m²

Numero Bar was built on a very narrow and long strip of land in the Jardins area, São Paulo. A walkway runs from the street through a hallway/tunnel fully covered in mirrors, leading to the main hall. At the entrance, the ceiling is extremely low and the view of the hall – cascaded – is unimpeded. Progressing towards the back, the height gradually increases: descending levels feature comfortable lounging areas, under a ceiling that extends on a continually rising surface. The low and indirect lighting throughout lends the ambience a pleasant and cozy atmosphere, which is perfect for a relaxing drink at the end of the day, accompanied by friends and the sound of good music. A room reserved for private functions takes up the lower floor and, contrary to the main hall, features a low ceiling all along, with couches placed in the central axis of the room, built-in overhead lighting throughout, and walls completely covered by antique posters or poster fragments.

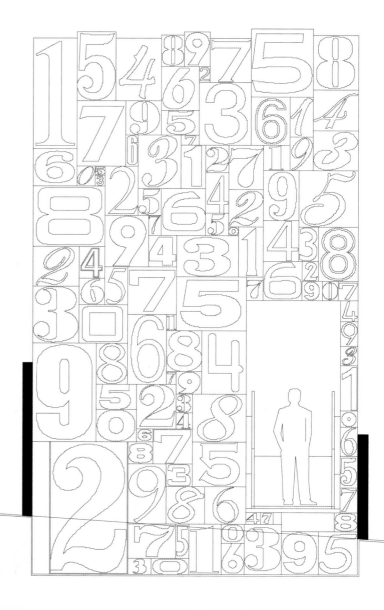

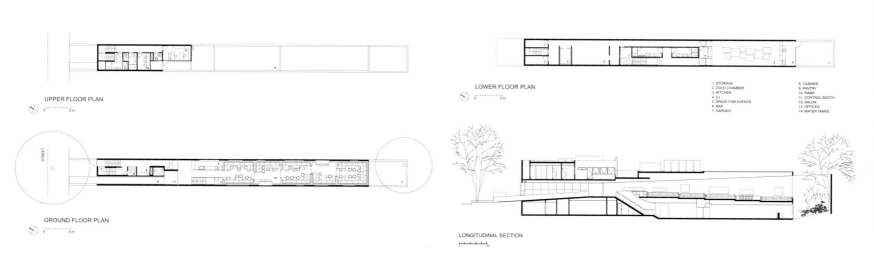

UPPER FLOOR PLAN

LOWER FLOOR PLAN

1. STORAGE
2. COLD CHAMBER
3. KITCHEN
4. DJ
5. SPACE FOR EVENTS
6. BAR
7. GARDEN
8. CASHIER
9. PANTRY
10. RAMP
11. CONTROL BOOTH
12. SALON
13. OFFICES
14. WATER TANKS

GROUND FLOOR PLAN

LONGITUDINAL SECTION

　　数字酒吧位于圣保罗市迦丁区一条又窄又长的地块上。

　　本案通道从大街穿过镶满镜子的走廊通往主厅。入口处的天花板非常低，沿着逐级降落的楼梯进入大厅后，视野则开阔起来。越往后走室内的高度越高：拾级而下就来到了舒适的休息区，天花板沿着一个逐渐上升的表面延伸开去。

　　四周低调的灯光给酒吧增添了一种愉悦舒适的氛围，当你忙碌了一整天，想和朋友们在美妙音乐中放松地喝上一杯时，这里将是你的完美之选。

　　下层是一间为私人宴会预留的房间，其特点与主厅相反，全低吊顶，内置顶灯，长沙发椅沿着房间中轴摆放，墙壁上贴满了古董海报或海报碎片。

La Folie

Designer: Jorge Luis Hernández Silva
Design Company: Hernandez Silva Architects
Location: Guadalajara, Jalisco, Mexico
Area: 750 m²
Photographer: Mito Covarrubias

The exterior is like a cone that catches and leads the visitor to a very narrow space opening again inside, accompanied by an undulating wall of segmented horizontal lines and bright colors that wraps around the room. It takes the adjacent spaces and joins the central space, forming a single large space. The project reuses the irregular structure as an important part of the concept. The whole space is a big box without irregular edges and separated by the building elements, all irregularities and multiple contractions are not covered. On the contrary, columns, walls, volumes changing on the floor are all covered in black, running and breathing freely. These are important elements from the concept. Even the setting of robotic lights, huge speakers and air vents are used for the project's decisions, forming a flat irregular expressive abstractions. The fact is that nothing gets in the way of the interpretation of design.

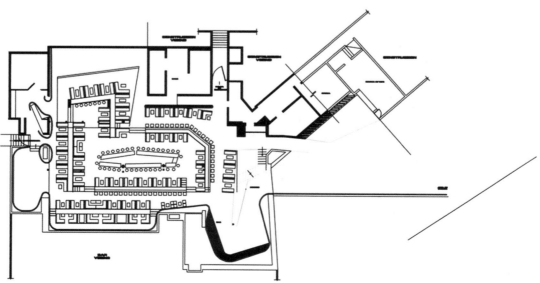

　　本案外形像一个圆锥体，引导客人从一个非常狭窄的空间走进里面开阔的空间，墙壁起起伏伏，室内被一段段水平的线条和鲜艳的色彩包裹着，四周分散的空间和中央部分浑然而成一个大空间。设计再度利用不规则结构作为其设计理念的重要组成部分，整个空间是一个没有边缘的、不规则的大盒子，并被内置元素分隔开，这样所有的不规则体和多倍压缩体可以毫无覆盖地自由流动。相反地，柱子、墙壁、地板上的变形体则被黑色所覆盖，自由地运行和呼吸，这些都是从设计理念而来的重要元素。就连那些自动照明灯、巨大的扬声器和通风口的安排和布置也被纳入本案的决策中，它们共同形成一个扁平的、不规则的、具有表现力的抽象概念。事实上，本案设计中没有任何障碍，空间理念完全得以诠释。

Library Bar

Designers: Cheryl Smith, Dexter Moren
Design Company: Dexter Moren Associates
Location: 12 Lancaster Gate, Bayswater, London, Uk
Photographers: Amy Murrell, Blue Pearl

The design concept by Dexter Moren Associates aims to rejuvenate the design and inject wit, charm and character into the building. The concept is based on creating a "country gentleman"s pied â terre' through the use of fabrics, pin stripes, herringbone, and elements such as the open fireplace and dark wood finishes.

As is often the case in the refurbishment of historic buildings, a significant proportion of work has been carried out on "behind the scene" upgrades, designed to bring it into the 21st century. The most significant attribute is the introduction of air conditioning and heating to provide a much more comfortable environment for guests. By executing the refurbishment programme over a series of phases the building has remained in operation throughout.

The Library Bar provides an eccentric feature and new heart to the whole project. The space has been reorganized and includes a new bar to provide an extended food & beverage offering. Accent lighting and contemporary furniture complete the setting.

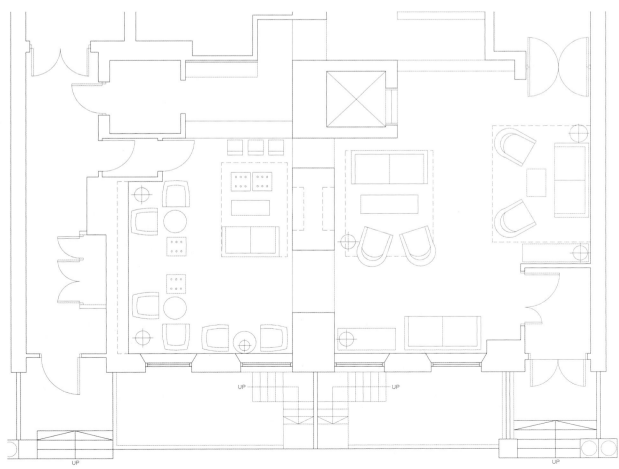

　　德克斯特·莫伦设计事物所的设计理念旨在使该项目重焕青春并为之注入更多智慧、魅力和个性元素。设计是本着打造一个"乡绅别墅"的初衷，大胆地使用纤维布料、细条纹布料、人字斜纹布料以及诸如开放式壁炉和黑木等装饰元素。

　　很多情况下，在历史建筑的重装中，很大一部分工作都放在了那些"背后元素"的提升上，从而使这里体现出21世纪的特征。

　　这种努力的最重要的标志就是装配空调和取暖系统，为客人提供更加舒适的居住环境。该项目的翻新工程历经几个阶段，而其经营运作却从始至终未有停止。

　　该项目为酒店增加了奇特的格调并注入了新的精神。室内空间经过重新改造，增设了一个吧间，可以为顾客提供更加丰富的食品和饮品。此外，局部照明和富有现代感的家具更为整体气氛锦上添花。

Monkey Bar Fumoir

Designers: Gian Frey, James Dyer-Smith
Design Company: Dyer-Smith & Frey
Location: Stüssihofstatt 3, CH-8001 Zürich, Switzerland
Area: 81 m²
Photographer: Beda Schmid

The Blue Monkey restaurant in Zurich, Niederdorf, braves its new smoking room and has indeed recognized the need to offer its guests a little more than just a musty room. Included in the entire concept created by the young Swiss designers, is the separate smoking room, equipped with a small, refined bar, a cozy lounge and seating made for comfort. Just as if the Monkey Bar had been set as a cigar lounge in the 1920s, it's been assigned classic, dark shades, and bright, highly set spotlights creating a welcoming, warm elegance.

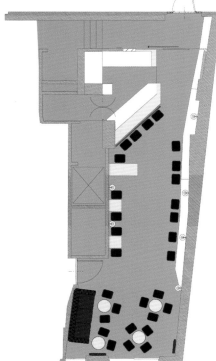

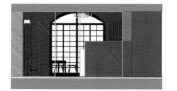

位于苏黎世尼德多夫的蓝猴餐厅大胆地开设了一间新的吸烟室,而且不仅是向顾客提供一个有香烟气味的房间。由年轻的瑞士设计师们提出的整个设计理念就是要有一个独立的吸烟室,里面配有一个小型的、格调高雅的吧台、一个温馨的休息室和舒适的坐椅。就像20世纪20年代的雪茄休闲吧的设计一样,这里也被赋予了古典的暗色调,而格调高贵的射灯更营造出了热情、温暖的优雅气氛。

Narcissus Bar Restaurant

Designer: Dimitris Naoumis
Loation: Neo Iraklio, M. Antypas street, Athens, Greece

Narcissus wishes to represent a new design perspective in the city of Athens. It is a relaxing place with selected, gourmet tastes, drinks and cocktails, which are combined with an elegant and classic, high aesthetics environment. Dimitris Naoumis creates a conceptual interior environment which tries to place the viewer in its very special surroundings.

The main aspect is to create a vintage styled place, in which, the main element is elegance. The highest quality materials have been used for the construction of the interior space. New and traditional materials such as wood, leather and granite are used in harmonic analogies in order to achieve functionality and style.

The proper lighting gives a scenographic feeling which is one of the main prospects of the designer. The lighting comes out of many different sources. The mixture of warm colors that dominate the place creates a positive, relaxing mood.

The big chandeliers are designed especially for Narcissus and they add an extra taste of elegance in the place. Most of the furniture and objects, are also designed especially for Narcissus by the designer.

The visual environment, contains several semiological references such as the cubistic version of a big classical wall painting behind the wooden bar. That makes a visual unity with the cubistic mirror which is between the two big chandeliers.

The vintage aesthetics of Narcissus and the gourmet tastes are combined with selected music that creates the proper background for chill out mood and entertainment. The place is proper for a coffee break, business meetings, entertainment, drinks and cocktails. Narcissus wishes to create positive and creative mood to the people who will decide to visit it.

位于雅典的水仙酒吧希望表现出一种新的设计视角。这是一个休闲之地，上等的美食、饮品和鸡尾酒与优雅经典、具有高度美学内涵的环境相映生辉。设计师创造出一种内部环境概念，试图将观者融入进来，从观者的视角进行设计。

设计的主要工作是创造一个风格高雅的场所，而优雅的格调应是主要元素。内部空间的装饰使用最高规格的材质。新型和传统的材料，如木材、皮革和大理石的运用都相得益彰，以实现实用功能与独特风格的平衡。

适当的灯光设计营造出透视效果，这是设计师主要想达到的目的之一。照明来自于不同的光源。酒吧以各种暖色灯光的搭配为主，创造出良好的休闲氛围。

那些大型的吊灯是专为水仙酒吧而设计的，为设计增添了格外高雅的格调。多数的家具和其他物品也是由设计师为水仙酒吧专门设计的。

酒吧内部环境的视觉效果包括几处有符号象征意义的作品，如木质吧台后面的立体主义的巨大古典壁画。这个设计在视觉上与两个大型吊灯之间的立体视觉镜面形成了统一。

水仙酒吧高雅的美学格调和美食与营造特定气氛的优美音乐一起，可以放松情绪、娱乐身心。它适合工余小憩、商务会谈、娱乐活动以及休闲进餐。水仙酒吧希望为每一位来客创造一种创新的积极氛围。

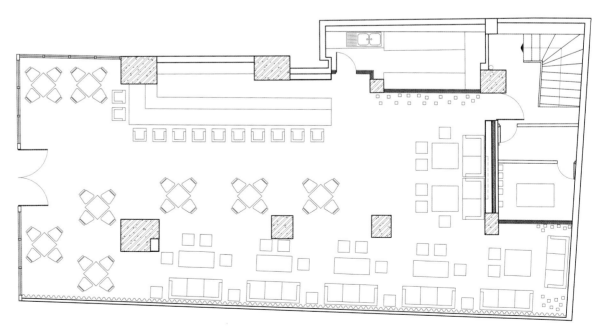

ΚΑΤΟΨΗ / ΚΛ. 1:50

Obikà Mozzarella Bar

Designer: Andrea Ottaviani
Design Company: Labics
Location: London, UK
Area: 250 m²
Photographer: Rob Parrish

The choice of placing Obikà Mozzarella Bar inside Northumberland House seems particularly correct if analyzed under an historical point of view or under the original role and function of the building. It is known, in fact, that the space described by the present report was originally conceived and designed to host the Restaurant/Banqueting Room for Hotel Victoria; seen under this point of view the Obikà design is a fundamental and interesting reinstatement of the original function. The design starts from the existing elements. It tries to find a new balance which bands together past and present.

The central morphology of the original space and the quality of the architectonical elements inside, have been the starting point of the design. In this sense, the new design is introduced inside the existing space in the "most respectful way", which means overlapping on it, without producing any variation.

The design is composed by the following elements: a big counter is placed in the centre of the space becoming a functional and visual point of reference for the whole space; for its function it gives the place a new centrality and finds an interesting dialogue with the existing columns; a complex wine-wall covers, like a sort of skin, the service area and a small part of the historic volume. In the last case it is not continued. It opens up. It has a dialogue with the existing space, stops in correspondence of decorations and architectonical elements, emphasizing their role and their presence. This new skin will be black unlike the white enamel historical one, asserting its temporal consistency through a material and chromatic difference.

如果考虑到历史因素或诺森伯兰府建造之时的初衷，将欧碧咖酒吧开在这里的确是个极正确的选择。按照时下报导所说，诺森伯兰府当初是为了给维多利亚酒店经营饭店或者宴会厅而设计兴建的。这么看来，欧碧咖是对这栋房子最初设计理念忠实而有趣的呈现。本案设计基于现有元素而又对其进行了新的解读，力求在历史与现状之间建立一种新的平衡。

本案以建筑原本格局和其内的建筑元素特质为出发点。从这一点来说，本案是对空间现状的最诚挚的致敬，仅对其进行修饰，而非添置任何变动。

酒吧设计由如下元素构成：空间中央放置了一个大型的吧台，作为整体空间功能和视觉的参照点；从功能上看，作为空间中心，吧台与原有的柱子产生有趣的对话效果；如同包柱效果一般，精致的红酒架包覆在柜台和柱子之上。它没有延续上一个设计的特色，而是开放式的。本案设计点与现有格局相互映衬，而非仅关注装饰性的元素，从而强调它们的功能和存在感。新的壁纸将会是黑色而非原来的白色釉面，通过材质和颜色来体现其延续性。

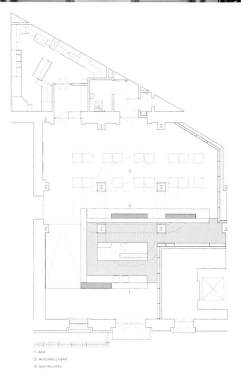

1. BAR
2. MOZZARELLA BAR
3. SEATING AREA

Soe

Design Company: gt2P
Location: Enjoy Santiago, Chile
Materials: Plywood 15mm, Acrylic 3 mm, Fittings of steel 2 mm, bolts and nuts 6.35 mm
Photographer: AryehKornfeld

Soe, or carob tree in mapudungún (Chilean native language), emulates the landscape of the typical trees and the foliage in the Valley of the Aconcagua, where this type of tree predominates.

Based on this idea, the rule or DNA "Section" is used to divide the volume resulting from the image of the trees. The subdivisions are determined in such a way that they exacerbate the continuous curvature of the surface of trees and leave irregular spaces which allow the light to pass through the foliage.

The structure is produced in plywood vertically sectioned with irregular distances and horizontal sections are settled perpendicular to the curve surface. From ceiling hang Pear App lamps of two colors that as a whole reinforce the forest concept.

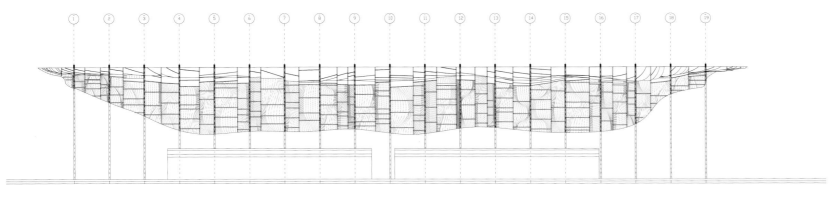

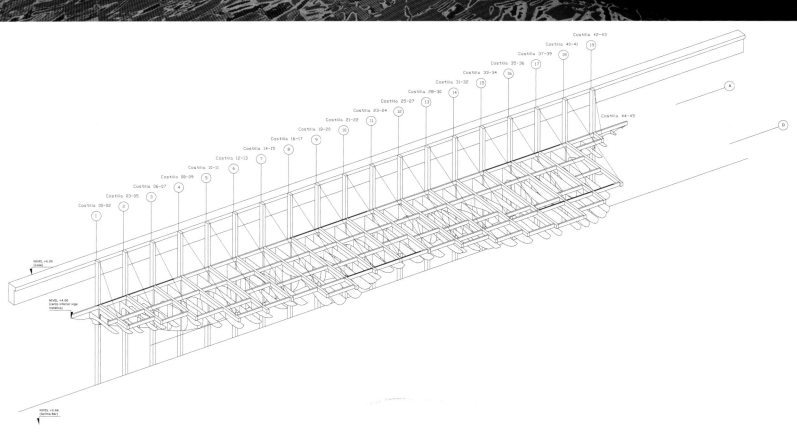

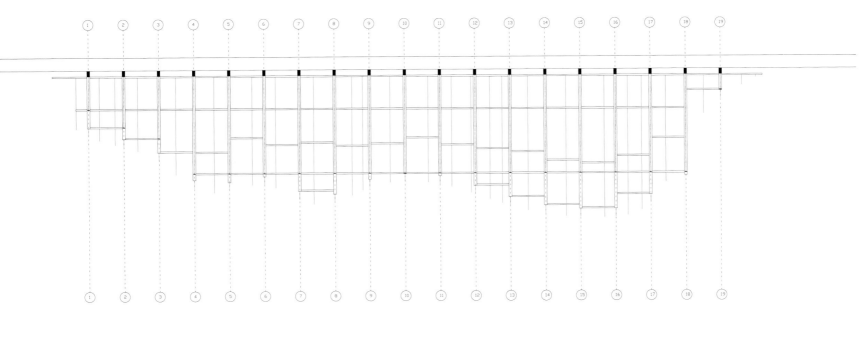

Soe在智利母语中意为角豆树，本案模仿阿空加瓜山谷典型树木和植物的形态，而角豆树广泛生长在该地区。

基于这种想法，设计师们用一定的规则或DNA"段"来划分树的形象所占的体量。这样的划分方式增大了树木表面的连续曲率并留出不规则的空间，使得光线可以照射进叶面。

整个结构由胶合板制成，采用不规则距离进行垂直分割，而水平段则垂直于曲面。

两种颜色的梨与苹果混搭造型灯悬挂于天花板下，从整体上加强营造出森林的效果。

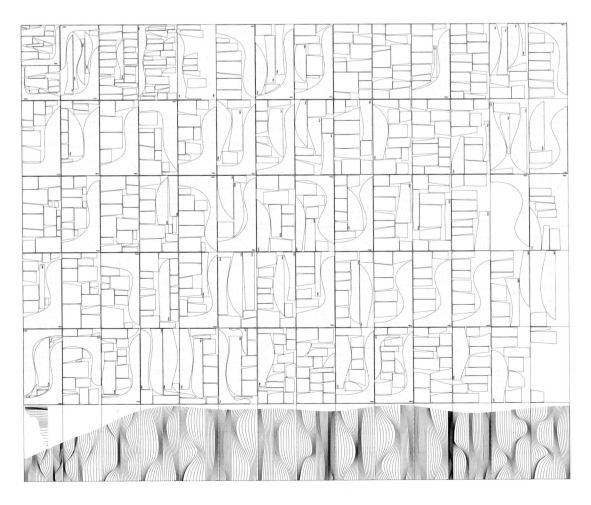

26 Lounge Bar

Designers: Miguel Rodenas, Jesús Olivares
Design Company : Cor & Partners
Location: Portal de Elche, Alicante, Spain
Area: 110 m²
Photographer: David Frutos

The proposal of the team was to create a "commercial and emotional window": a place where looking and being looked beneath the huge ficus on the "portal", which would generate a stable emotional link that made customers, come back.

The designers propose the construction of a "controlled interior volume" able to suit lighting, temperature and acoustic sensation. Built with two faced "u"s: one made of industrial timber that fits on the floor, at the bar and forms the bench going up to 1.8 meters height, and another punched "u" completely white, descending from the ceiling where all facilities (air conditioning and sound) are accommodated.

These "u"s are crossed in a weightless way by two adjustable light sets that bathe the room from its center, and try to emphasize and define the interior space, letting perceive "a box within a very old local".

It is a phenomenological project. The states of light are very important due to temporary uses and change on supplies depending on the time of the day. Say it is not an area that changes depending on the time, but rather different premises, where different scenarios are set, almost like a stage. Ground lighting takes all its power at night, diminishing its intensity wall washers and the two great lamps, getting a room similar to the bistros of W. Allen. In contrast, in the morning are the two great lamps that catch the spotlight, making appear clearly defined spatial boundaries, something that reminds the photography of Slawomir Idziak.

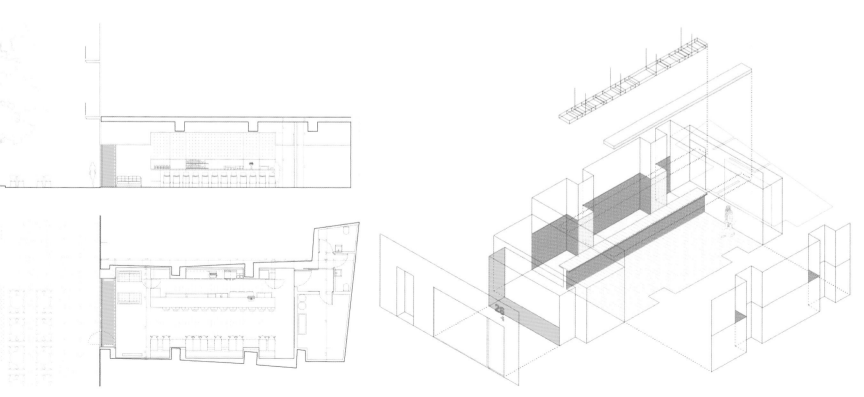

设计团队提议创建一个带有"商业和情感窗口"的酒吧：站在入口，可以看到一棵巨大的榕树与之遥相辉映，这将产生一个稳定的情感纽带，牵引着顾客再次到此光顾。

设计师们欲建造一个"可控制的内部空间"，以此满足照明、温度和音响效果的要求。设计概念由两个双面的字母"u"组成：其中一面是用工业木材制成，适合酒吧的地板使用，组装的椅子则高达1.8米；另一面则完全呈白色，从安装有所有设施（空调和音响设备）的天花板延伸而下。

这两个"u"以失重的方式相互交错，通过两个可调节的照明设备从中心开始将灯光洒向室内周围，试着凸显和界定内部空间，宛如"一个非常古老酒店中的盒子"。

这是一个基于现象学的项目。由于需要临时光源，并且随不同时段改变光源，光的状态变得非常重要。如果说这不是一个随时间而变的场所的话，那么可以说是在不同的前提下，变换着不同的场景，宛如一个舞台。由于晚上地面照明发出强大的光源，减弱了洗墙灯和两个巨大灯体发出的光线，使之呈现的是一间类似小酒馆似的房间。相比之下，清晨两个巨大灯体吸引着公众的注意力，出现界限清晰的空间边缘，让人们不由地想起斯拉沃米尔·埃迪扎克的摄影作品。

Q Bar

Design Company: crossboundaries architects
Structure and Material: steel structure, fibre-cement panel, wood cladding
Location: Beijing, China
Area: 460 m²
Photographers: Yang Chao Ying, Jeff Hinson

The designers were approached by the owners of Q Bar to design the new rooftop.

The design was required to solve several issues: to contain the sound produced by the clientele, to create a half-enclosed environment, with different views, with places to partially hide.

The designers thought that the best solution in this case was to use an architectural device that could partially enclose the space. The abstract vertical stele almost as a sculpture is 3.5 meters high. By duplicating this device on the rooftop the space is organized, separated, and fragmented. The result is an environment that allows intimacy and stimulates curiosity.

A second device is the bed-box. The initial idea was to frame the night view of the city. The box provides shelter and intimacy and it is also a practical storage during the winter season. The interior space of Q Bar has an interesting concept: nothing is on the walls. Only the red paint that creates a plain background for the bar crowd. In the same way the outdoor rooftop uses the same red combined with the wooden panels. Only these two materials are used in order to reduce the visual pollution.

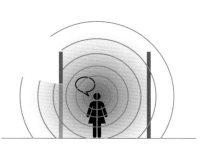

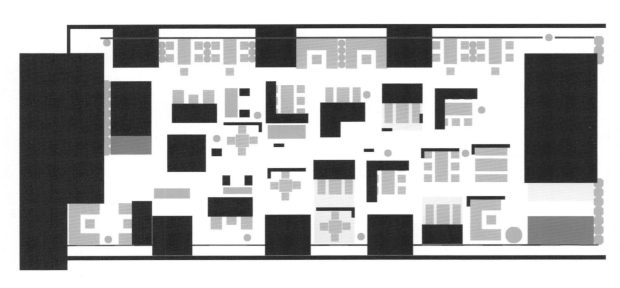
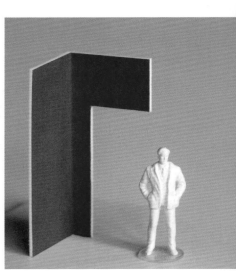

当初Q吧的业主要求设计师为其设计新的天台。

设计要求是要解决几个问题：为顾客隔音、设计半封闭的、体现不同创意的空间景观。

设计师们认为本案最好的解决途径就是使用构架结构来实现对象空间的横向或纵向的半封闭。所以设计了一根3.5米高的抽象立柱，形如一座雕塑。通过这个方法将天台实现了整合，并分割成独立的空间。此种设计的效果是营造了私密的空间，并使顾客生起探究的兴趣。

第二个方法是箱形座架。设计师们最初的想法是这样的设计可使顾客欣赏到城市夜景。该设计既可遮风挡雨也可提供私密空间，在冬季还可用做储藏空间。Q吧的内部空间颇为有趣：墙壁上空空如也，只有一层红漆为客人提供朴素的背景。室外的屋顶天台也使用了同样的红色，并使用了木制嵌板。整个设计只使用了这两种材质以减少视觉上的杂乱之感。

Simyone Lounge

Designers: Lionel Ohayon, Siobhan Barry
Design Company: ICRAVE
Area: 204 m²

Simyone Lounge was designed to be a wall to wall party place where the walls of the space literally create the energy and set the mood. In the front bar room, frames are intentionally left empty to showcase the raw, stripped down beauty of the masonry walls behind. Highly reflective ceilings are designed to maximize the height and to give a mirroring effect, where the clientele actually become the fabric of the space.

The back room was designed to reflect the high energy revellers of the neighborhood. A sculptural faceted wall, with a warm rich palette of perforated ebony stained wood, frosted glass and mirror encapsulates the space. During the early lounge hours, a warm amber glow emanates from the walls to create an amorous ambiance. As the night progresses and the energy picks up, the walls begin to dance,Pulsate and chase around.

Lighting elements bring the space from a low key romantic space to a high energy lounge.

　　这里被设计成拥挤的派对风格，在这儿，整个空间的墙壁真正营造出充满活力的氛围。在酒吧前厅的墙上，刻意空置的画框突出了背景石墙的原始低调的美感。反射效果极佳的天花板最大限度地从视觉上拉高房间高度，并利用镜面效果使得顾客本身也成了空间装饰的一部分。

　　后厅与激情四射的前厅产生强烈对比。雕花切面的墙壁以暖色系乌木镂空雕刻、磨砂玻璃和镜子为装饰物，奠定了空间的基调。傍晚休闲时，墙面上映射的琥珀色光线映照出温馨的氛围。随着夜色渐浓，温馨被激情取代，墙壁也开始跃动起来，光线犹如跳舞般相互追逐。光线元素使整个空间从低调浪漫变得活力四射。

Villa

Designers: Lionel Ohayon, Siobhan Barry
Design Company: ICRAVE
Location: West Hollywood, California, USA
Area: 167 m²

"The general idea is that it's a residential house, says Villa's lead designer and architect, Lionel Ohayon, "but an eccentric Hollywood Hills home. Things seem very familiar, a little off-center at the same time."

Ohayon isn't joking. Upon entering, one of the first things attracting notice above the bar is a bird cage large enough to accommodate a burlesque dancer, flanked by a NASA spacesuit and a stuffed peacock (a stuffed bison head was replaced in January to keep the lounge's "fresh"). "Those are two different kinds of interesting 'suits,'" Ohayon says of the peacock and the so-called moon man."We are playing those juxtaposing ideas."

The high-concept interior (replete with a stunning hanging rope canopy meant to emulate a chandelier, and backlighted books in the library) might be lost on some of Villa's audience, however. With the tightest door in town (it's nearly impossible to get into the bar unless you are on the list), the crowd tends to be heavy on actresses, hospitality industry insiders and the occasional agent or two. There's no VIP area because the whole place is a VIP area.

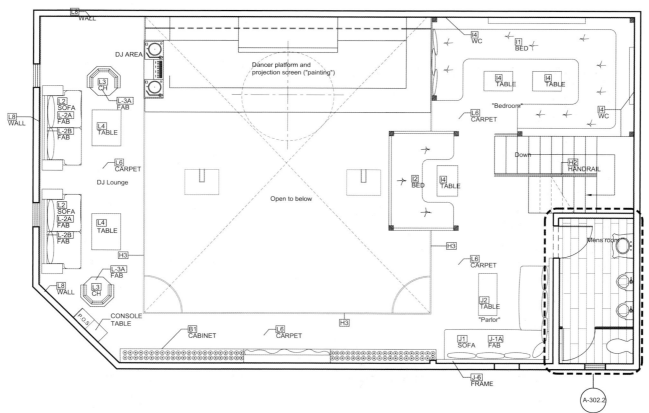

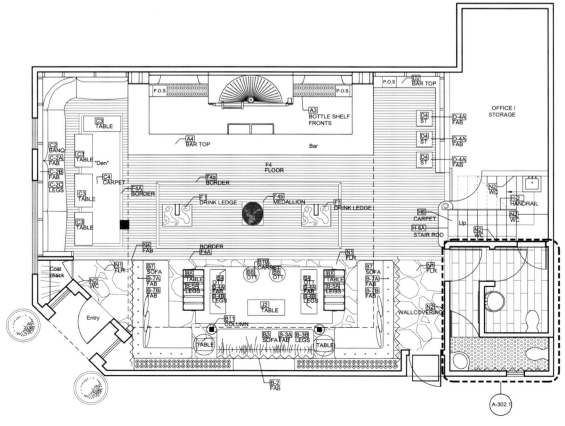

  本项目的首席设计师和建筑师莱昂纳尔·欧恩说:"设计的大体的想法是把这儿当做一所居住的房子,但是是一所异乎寻常的位于好莱坞山的房子。一切看似熟悉但是一切又那么不中规中矩。"

  欧恩并没开玩笑。一走进酒吧,客人们首先注意到的是在酒吧上方有一个巨大的鸟笼,足够容纳一个舞者,而它的两边分别放了一件美国国家航空航天局的太空服和一个孔雀标本(起初是个牛头,为了保持酒吧的新鲜感,在一月份的时候它被替换了下来)。"它们是两种截然不同而又有趣的服装"当欧恩说起这个孔雀和所谓的"登月人"时说,"我们想玩转一种混搭的主题"。

  相当具有概念性的内部设施(到处都是令人惊叹的装饰品,比方说吊灯状的高空悬绳和有背光的图书)可能会使别墅风情流失一部分客人。但是作为城中一家并非面向所有人开放的酒吧(若非在受邀客人名单中,你几乎不可能进场),它的客人大多是女演员、酒店业内部人士或偶尔一两个经纪人。酒店没有设贵宾区,因为整家酒吧都是贵宾区。

XVI Lounge

Designer: Francois Frossard
Design Company: Francois Frossard Design
Location: New York, USA
Area: 232 m²

While the rooftop lounge nestled into the corner of 48th and 8th avenues in Manhattan has had a few prior incarnations, none has had such a rooted backstory and well-thought-out concept as its current occupant. XVI, the roman numerals correspond to the number 16, which is the floor which the venue is on. The opulent décor designed by interior designer Francois Frossard, renowned for his Louis the XVI style of furnishing rooms with a modern twist, is one of New York's trendiest spots. Modeled after the estate and lavish lifestyle of King Louis the 16th, since Louis the XVI was known for his opulent reign, which only came to a halt when he was beheaded, XVI is dubbing the venue "Versailles in the Sky" and the logo is the guillotine. The golden accents, rugs and seating are all reminiscent of France's most luxurious and infamous period. High-backed chairs and couches dot the room, while flourishes such as modern sculptures and lanterns — home to five fires — are spread out over the lounge. At about 232.26 square meters, approximately 260 people can enjoy one of the 25 tables or free space.

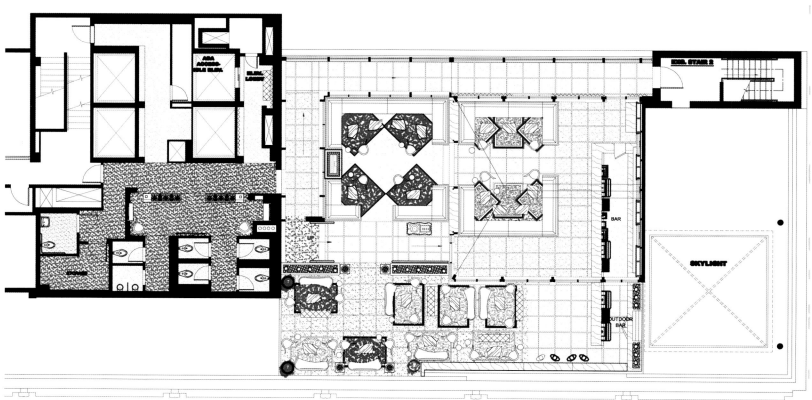

FLOOR PLAN 16th Floor

本案地理位置优越，在曼哈顿第四十八大街和第八大街交汇处的天台之上。从前位于此处的酒吧都不具备如今"十六号酒廊"这样深厚的历史背景和深思熟虑的设计理念。罗马数字XVI对应数字16，因为酒廊在第16层。本案华丽的设计出自室内设计师弗朗索瓦·弗罗萨德之手，他把现代元素完美地融合在路易十六风格的装饰里，闻名遐迩，是纽约时尚焦点之一。本案设计以路易十六国王象征着地位和财富的生活方式为范本。路易十六因其统治期间的富庶而闻名，直到被斩首才宣告结束。十六号酒廊的室内配乐是"空中凡尔赛"，而酒廊标志则是断头台。地毯和座位的金色镶边让人想起法国历史上那段最奢华而臭名昭著的时代。高背椅和长沙发点缀着房间，酒廊里随处可见现代雕塑和灯笼等华丽的装饰物。酒廊占地约232.26平方米，可容纳约260位客人落座于25张餐桌，或在此畅享一片自由空间。

Starlite

Designers: Jason Lane, Barbara Rourke, Jason St. John
Design Company: Bell & Whistles
Location: San Diego, California, USA
Area: 251 m²
Photographer: Several

Starlite is one of San Diego's most custom interiors with the designers / builders having their hand in all aspects of creating the space. A smoke tinted glass door opens up to an slatted hexagonal hallway which leads directly to Starlite's sunken white bar. Above the bar hangs a striking chandelier crafted from varied lengths of stainless steel tubing, angled at the ends to reveal the embedded lights to twinkle and reflect in the mirrored ceiling. Natural elements such as stacked stone, walnut paneling, and cork tiles, compliment the gold pendants and luxurious black leather booths.

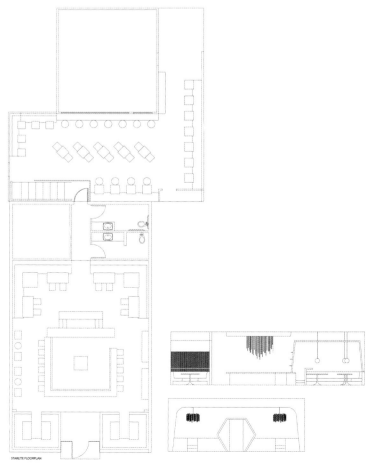

STARLITE FLOORPLAN

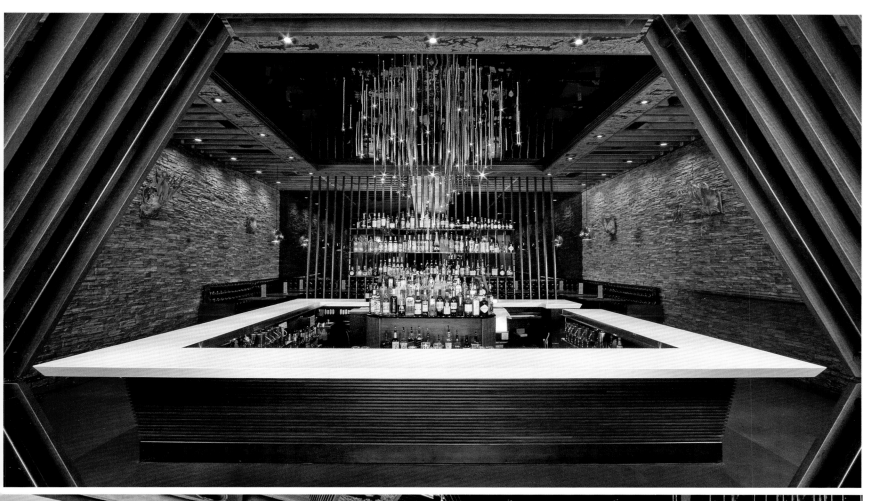

　　星光酒吧是圣迭戈众多定制的内部空间之一，设计者/建筑师参与了整个空间的构造过程，可谓事无巨细。打开一扇烟雾茶色玻璃门，穿过一条用木板条构成的六边形门廊，便可以直达里面的白色吧台。吧台上方悬挂着一盏醒目的吊灯，它由不同长度的不锈钢管制成，末端呈一定角度，以呈现内置灯的光线，并在天花板上折射出来。自然元素，如叠石桥、胡桃木镶板和软木砖等，同金色吊饰和豪华黑色真皮包间融为一体，相得益彰。

Mouton Cadet Wine Bar

Design Company: Naço Architectures
Location: No. 9 Jiansheliu Road, Yuexiu District, Guangzhou, China

Understanding precisely the values and brand DNA, Naço Architectures has developed a design concept that explores and enhances the wine brand.

The Mouton Cadet bar is an invitation to travel. The design concept tells the story of a journey and shows the brand international dimension.

All the elements of the bar recalls exceptional moments or sensations a dandy traveller has experienced, and the unexpected atmospheres or refined places he has known or been to and is seeking for: travel cases displayed on the wall and floor, travel art work, wooden floor, brass element, and velvet fabric.

Naço Architectures was able to design a modern and elegant bar, impressive of Mouton Cadet wine brands. In the open small bar area, a selection of refine stone and brass material wrapped around the bar enhances the elegance of Mouton Cadet brand. The bar area overlooks an outdoor external terrace where tall bamboos give a harmony ambiance. The open lounge sofa area offers wine culture and journey visual displays. In both areas, mirrors with Mouton Cadet sheep icon pattern are etched and sandblasted to the suspended ceiling.

Mouton Cadet also incorporates a VIP space for exclusive Rothschild wine collection. Baron Philippe de Rothschild being the world's most exclusive legendary wine brand in the wine universe, the design space aims to be exclusively premium and sophisticated. This leather lounge sofa seating area is a place where customer can relax and enjoy the sensational taste of Rothschild Wines. It is a comfortable and intimate place where the design is mature and the furniture are classical. Here, every material details: the copper-metal textures, stone flooring, the wood paneling enhance the exclusivity and high-end of the Rothschild wine collection.

Mouton Cadet Wine Bar not only recalls a travelling experience, it is also a journey to the heart of wine culture and elegance.

在精准把握木桐嘉棣红酒吧的品牌价值后,纳索事务所形成了酒吧的设计理念,开拓并提升了酒吧的品牌价值。

酒吧的设计理念源于对旅行的邀请。旅行中的故事被婉婉道来,彰显了该品牌的国际影响。

墙上、地板上陈列的旅行箱、艺术画、木质地板、铜器、天鹅绒纺织品等元素,能触发一个时尚旅人对旅途中特殊时刻和情感的回忆,也能使他们想起那些他们已知的、曾到访过的或是正在寻觅的别致氛围和精致场所。

纳索事务所设计了一个现代而又优雅的酒吧,体现了木桐嘉棣红酒品牌独具一格的吸引力。开放式酒吧区,以精选石材和铜质材料进行包裹装饰,凸显了木桐嘉棣的优雅气质。酒吧区可俯瞰室外露台,那里种有高挑的竹子,营造了一种和谐的氛围。开放式的大堂沙发可使顾客体验红酒文化和视觉享受。这两个区域的天花板上都刻有该品牌羊形标志的镜面装饰。

木桐嘉棣酒吧还设有贵宾间,特供罗思柴尔德(拉菲)系列红酒精选。菲利普·罗思柴尔德男爵系列是全球独一无二的红酒传奇品牌,因此贵宾间设计也以高端优雅为目标。皮质休闲沙发坐区是客人们可以放松并享受罗思柴尔德红酒绝妙风味的地方。这是一个舒适而私密的场所,设计风格成熟,装饰家具经典。在这儿,所有的细节:铜质金属肌理、石质地板、实木嵌板,都烘托了该系列红酒的高端贵族品质。

木桐嘉棣酒吧不仅唤起人们对旅途经历的回忆,也是一次通往优雅红酒文化精髓的旅程。

Buck and Breck

Designer: Ingo Strobel
Design Company: Motorberlin
Project Architecture: C+ Architects
Location: Mitte District, Berlin, Germany
Photographer: Katja Hiendlmayer

Looking from the street the only sign for a business at the location is a huge and well lighted gallery window with changing installations curated by the artist Theo Ligthart. After entering the unmarked dark door next to it, the guest walks through a narrow corridor, passing two corners. The guest room is dominated by a huge table, surrounded by the bar stools.

The table actually is the bar, lacking the vertical step which normally at a bar separates the guest area from the working area. A further detail which enhances the intimacy of the place is the fact, that guests are also seated "behind the bar", on the barkeeper's side of the table.

Under the ceiling around the walls of the main room runs a belt of drywalling which houses indirect lighting, and is covered with a fresko showing poisonous plants in the dark. The reflecting bays in the walls are colored gold, as well as the insides of the huge lamps under the ceiling. This arrangement of ceiling lights is the only other remarkable object next to the bar, giving the huge table a certain counterweight.

The Buck and Breck is a wonderful example for the creation of an intimate atmosphere for connoisseurs of quality cocktails. It enables guests and barkeepers to meet close & personal. Gonçalo de Sousa Monteiro and his companion Holger Groll have developed an individual mixing style, defining quality in a unique way. The menue is a collection of minimalistic variations of well known cocktails and forgotten classics from the dawn of the cocktail era. The bar is expected to give this experience a stage, and provide a calm intimate retreat for the guest.

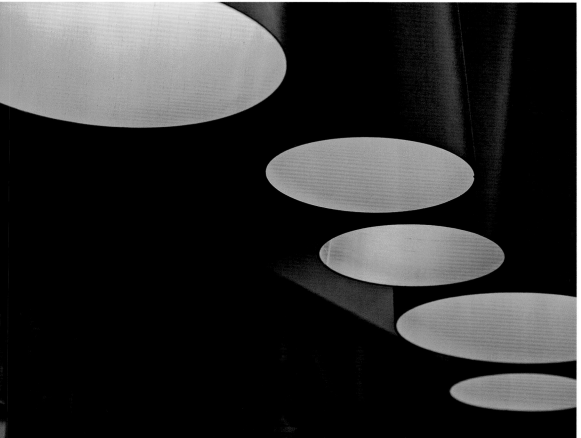

　　临街望去，巴克与布雷克酒吧的唯一标志是一个光线充足的巨大艺术窗，里面是由艺术家西奥·莱萨特设计的、不断变化的装置。走进旁边那扇隐秘的暗门，客人们进入到一条狭窄的走廊，沿走廊转两个弯到达客房，客房里安置着一张巨大的桌子和一圈酒吧凳。

　　实际上这张大桌就是吧台，这里不像其他酒吧，没有纵向地把顾客区和工作区划分开来。另外一个细节在于顾客也坐在"吧台后面"，和酒保一样坐在桌子的同一侧，亲密感因此而倍增。

　　在天花板之下，干式墙环绕主室的四周，里面隐约透着光亮，表面饰以壁画，壁画图案为黑暗中的有毒植物。墙壁里反射光的突起部分和天花板下面的巨型吊灯内部都涂成了金色。天花板下的这些灯是吧台旁边唯一一个引人注目的设计，用以在视觉上平衡那张巨大的吧台。

　　巴克与布雷克酒吧的设计为高品质鸡尾酒的鉴赏家们创造出一种亲密的氛围。客人和酒保可在此亲近而隐秘地会面。古赛鲁·德索萨·蒙泰罗和其同伴霍尔格·格罗尔开创了一种独特的混合风格，以独特的方式诠释高品质。酒单囊括了著名鸡尾酒的各种极简化种类以及鸡尾酒时代伊始时那些被遗忘的经典品种。巴克与布雷克酒吧为这种体验提供一个平台，客人可来此享受安静、私密的空间。

Eighty-Six

Design Company: designLSM

The design intent was to create a bar and restaurant which exuded quality and discreet glamour with an avant garde art deco twist. This was achieved by: creating more space for the ground floor front of house by removing the large island bar and relocating it to the left hand side of the room.

Designing a grand feature of angled gilt mirrors fixed to ceiling of the ground and mezzanine floors; this emphasizes the visual interaction between the two spaces, creates different perspectives around the rooms and gives the impression of ceiling height to the mezzanine.

The use of quality materials such as the beautifully finished mirror polished steel cocktail bar, charcoal velvet chesterfields with chrome polished buttons and floors of dark wood and slate.

The walls on the ground floor are black lacquered random sized panels while the restaurant has a more informal feel with a backdrop of glamorous gold panels cladding the walls, which is also adorned with original artwork by Charlotte Cory.

Artwork in gilded frames, a grand dining table, dark wood paneling and wine displays create a luxurious private dining area.

The success of the design is that it has set Eighty-Six apart from its other Fulham and Chelsea competitors, creating a destination hotspot for the well-heeled diners of the area. The overall quality of the setting and surroundings are sufficiently alluring to attract visitors from considerably further than SW postcodes whether it be for an informal drink, evening meal or special occasion. The new venue design has proved popular with ciritics and has appeared in two TV programmes ("Made in Chelsea" and "Young, Rich and House Hunting").

Marina O'Loughlin at the Metro quotes in her review: "the new look, from designLSM, the team behind the striking Galvin La Chapelle, is hot stuff – I love the steampunk-esque animal-headed portraits, starlet light bulbs and gilded-mirror ceiling."

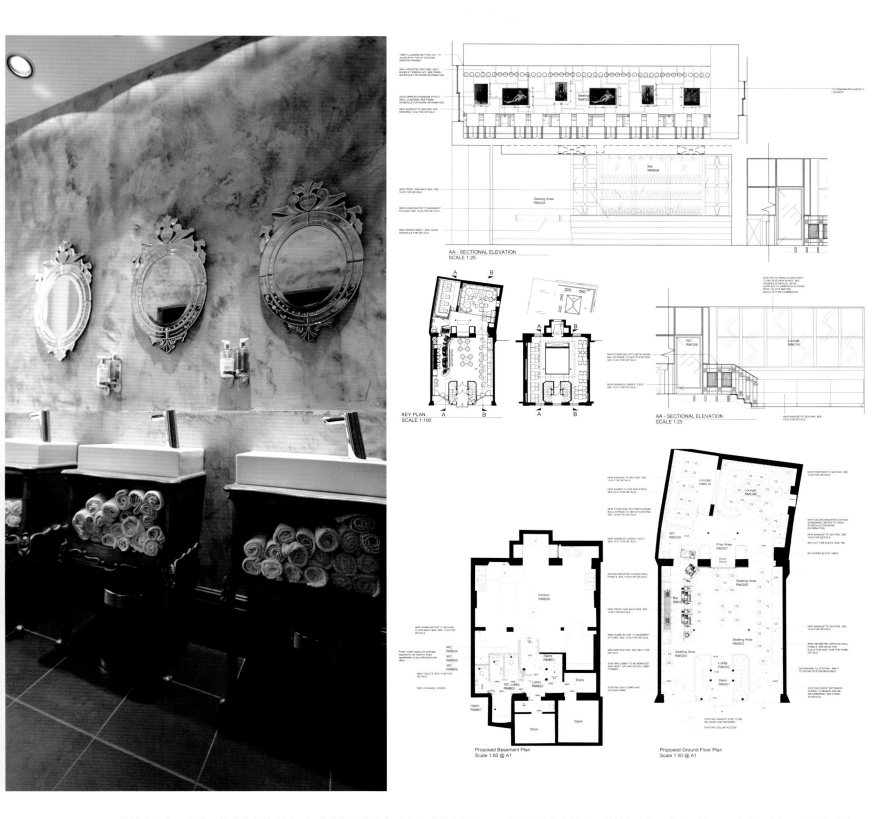

设计者旨在运用前卫的艺术装饰创建一间处处彰显高贵气质和细致魅力的酒吧。这是通过在房前为一楼扩充空间、将大型内置吧台移至房间左侧的方式实现的。

一楼及夹层地面安装的侧角镀金镜极具特色，它加强了两个空间之间的视觉互动，营造出房间内不同的观赏视角，给人一种天花板高度延伸至夹层中的印象。

酒吧装饰使用优质的材料，如精美的镜面、抛光钢质鸡尾酒吧台、灰色天鹅绒长款沙发、配有铬合金打磨的钮扣以及铺设的黑木和石板地板。

一楼墙壁镶嵌着黑漆涂制而成的规格不一的壁板，使餐厅的感觉更随性一些，背景是迷人的金色壁板铺设的墙壁，同样辅之以夏洛特·科里的原创艺术品。

镀金框架里陈设的艺术品，一个豪华餐桌、黑木壁板和葡萄酒展区，这些营造了一个具有奢华氛围的私人用餐空间。

设计的成功之处在于同富勒姆和切尔西等竞争对手相比，86酒吧成为该区有品位的用餐者的首选之地。总而言之，无论是一场非正式的酒宴、晚宴，还是为特殊场合举行的宴会，86酒吧的选址及优雅的周边环境吸引着远至伦敦西南的用餐者驱车前来。新的场景设计获得了评论家的好评，并在两档电视节目中出镜。

玛丽娜·奥-洛林在餐饮评论专栏中提到："新面貌，出自设计LSM事务所之手，这支设计了备受瞩目的高尔文·拉·沙佩勒餐厅的团队，是了不起的；我喜爱蒸汽朋克式的动物头像、星光点点的灯泡和镀金镜面天花板。"

Asphalt

Design Company: karhard architektur + design
Location: Berlin, Germany
Photographers: Friederike von Rauch, Stefan Wolf Lucks, Henrik Jordan

The basement of the Hilton Hotel at Gendarmenmarkt, Berlin (Mitte), formerly known as Trader Vics, has been transformed into a modern space complete with bar, lounges, dancefloor and a live music venue. All existing fixtures were removed to reveal the buildings skeleton so as to measure accurately the acoustical properties of the blank space. To tackle the soundproofing, the perimeter walls of the entire club were equipped with floating partition walls and the freestanding structural columns were clad in brickwork, which served to decisively shape the overall appearance of the design.

A key element of the design is the centrally arranged stage which serves to divide the overall space into a club area with a dance floor and a bar area with a live music platform.

The furniture concept was to use a balance of both used and new by selecting contemporary designer elements as well as lesser known pieces such as salvaged Gymnastic Vaulting Boxes and Vintage Mirrors.

Key points:
- The flooring of colored polished asphalt, became the namesake for the club.
- Lining the walls with a woven anodized steel mesh to improve room acoustics.
- Riveted brass panelled sliding doors to subdivide the spaces.
- Fanlights of perforated brass with LED technology.
- Parapet and bar panelling made of rough, dark stained timber.
- Leather seating upholstery in Aubergine, Mocca and Olive Green back rest. Textiles manufactured by Romo.
- Extensive worldwide furniture hunt.
- Sound system over the dance floor as a combined system of array loudspeakers with attached prefabricated sound ceiling with over 350 loudspeakers and built-in lighting with halogen spotlights.

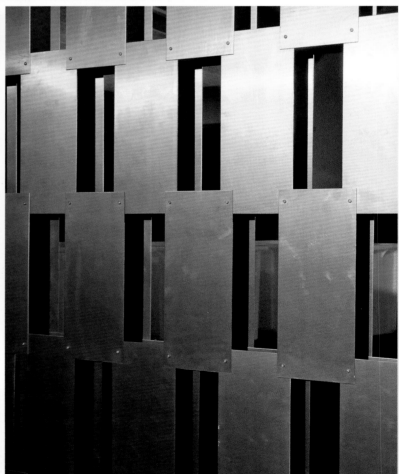

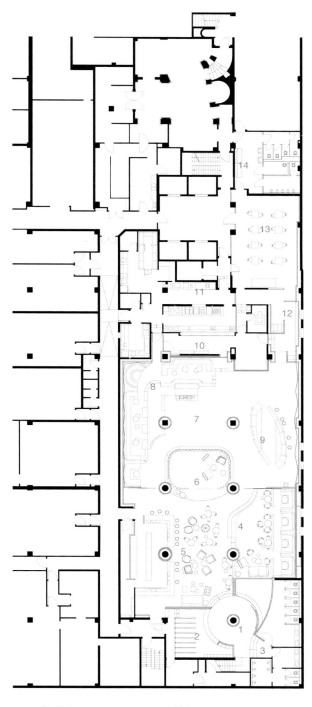

01 Entrance
02 Wardrobe
03 WC
04 Lounge/Restaurant
05 Bar
06 Stage
07 Dance floor
08 Lounge
09 Bar
10 Raucher lounge
11 Kitchen
12 Passageway
13 Restaurant/Club
14 WC

　　沥青酒吧位于柏林中央区宪兵广场的希尔顿酒店的地下室，原名为垂德维客，现已被改造成一个现代化的空间，酒吧、休息室、舞池及现场音乐表演场地一应俱全。

　　拆除了现有的固定装置，建筑框架一览无余，这样可以对空旷的空间所具有的声学特性进行准确测量。

　　为了解决隔音问题，整个设计的周边墙体都配有浮动的隔音墙，独立结构的柱状体上覆盖着砖块，这决定着设计的整体外观。

　　在酒吧设计中起决定性作用的是位于中央的舞台，它将整个空间分成了一个配有舞池的会所和一个配有现场音乐表演舞台的酒吧区。

　　家具设计旨在于陈旧的和崭新的家具中取得一种平衡，精选当代设计元素，并运用一些鲜为人知的家具，如体操跳马箱和古董镜。

设计要点：

　　——酒吧得名的彩色抛光沥青地板。
　　——墙体布满的电镀钢丝网以改善室内空间的声音效果。
　　——黄铜镶板推拉门将空间再分割。
　　——采用LED技术的黄铜扇形窗。
　　——由木质粗糙，涂黑的木材做成的栏杆和酒吧镶板。
　　——紫红色、摩卡色和橄榄绿靠背的皮革座垫及罗莫制造的纺织品。
　　——在全球范围内对家具进行广泛筛选。
　　——舞池上的音响系统是一系列扬声器的组合，配有350个扬声器的人工制成的音响天花板和内嵌式卤素射灯照明。

Tribeca Grand Hotel Lounge

Designer: Winka Dubbeldam
Design Company: Archi-Tectonics
Team: Pauline Magnusson
Location: Tribeca Grand Hotel, New york, USA
Contractor: Galcon Enterprise Inc
Photographer: Floto & Warner

The new Lounge in the Tribeca Grand Hotel was curated by Richard Klein of Surface Magazine. The designers were invited to design the space and to incorporate with 18 of the best modern Spanish designers and manufacturers. When researching their work it struck the designers that one of the companies, BD Barcelona, represents amazing Dali furniture. Dalí, of course, was known for the striking and bizarre images in his surrealist work, as shown below in the The Persistence of Memory, a painting which epitomizes Dalí's theory of "softness" and "hardness", which was central to his thinking at the time. Many also consider that the melting watches were there to symbolize the irrelevance of time, which, in the Tribeca Lounge, is of no relevance either.

The "Softness" expresses itself in the colors, the "melted" forms and the fluidity of the compositions, which became the main themes for the space. The centerpiece, a group of the Dalilips sofas, became the pivot point around which the space organizes itself. The Dalilip sofa, is made in the shape of a mouth which the artist created together with Oscar Tusquets in 1972 for the Mae West room at the Dalí Museum in Figueres, Spain.

The "Hardness", expresses itself in the bar area, where the wall behind the bar was transformed to become a glass vitrine for figurines made by LLadro. Their slightly surrealist character was further expressed by putting them in this translucent state, where their white and gold colors are soft glittering suspended moments in an otherwise dark wood wall. The Bar itself is designed by BD Barcelona, its stark shape is contrasted by the legs, each one a sculpture in itself.

The overall theme for the Lounge was a relaxed, luxurious, and dark glowy atmosphere, hence the designers selected these works or color, shape, and surrealist aspects (sofa's with inserted lamps and tables, vases with faces, low tables with hollow shapes inserted etc). An important aspect in the design was that the selection of luxurious wallpapers, soft lighting, metallic curtains and screens, and furniture would designate specific areas, such as a bar area in dark plum leather, a lounging area in dark intimate colors, and the Dalilip sofa area with a deep red cloud of suspended lights above. All this, to result in a coherent spatial environment for the new Salon.

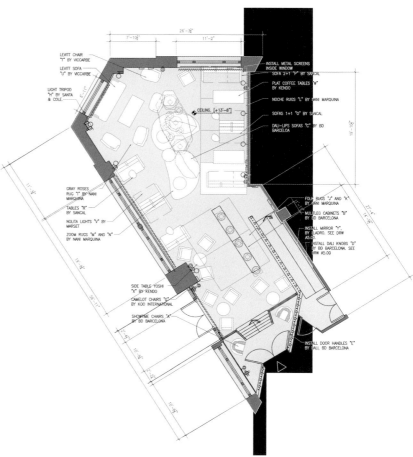

  特里贝加大饭店新的休闲吧由《平面杂志》的理查德·克莱恩掌管。设计师们受邀设计本案时，要与18家顶级的西班牙设计事务所和生产厂家通力合作。在调研他们的作品时设计师们惊讶地发现，其中一家巴塞罗那BD公司有着令人称奇的达利家具风格。当然，达利以惊艳怪诞的超现实主义著称，如他的作品《记忆的永恒》所表现的那样。这幅作品集中体现了达利的"至刚"和"至柔"的理论，这是他当时重要的思维方式。很多人也认为那些融化的钟表代表着时间的不相关性。在特里贝加休闲吧，时间也不具有任何关联。

  "至柔"特点表现在颜色、"融化"的造型和各种构成元素的流动形状等方面，这些组成了该休闲吧的主要主题。空间的中心装饰是一组达利"唇形沙发"，其他装饰都围绕这个中心点布局。顾名思义，达利"唇形沙发"设计成了嘴唇的形状，这是艺术家同奥斯卡·图斯凯茨在1972年为达利博物馆的梅西厅设计的，该博物馆位于西班牙的菲格雷斯。

  "至刚"表现在吧台区，在这里，吧台后面的墙壁被改造成了玻璃陈列柜，用以展示拉德罗的雕像作品，这些作品的超现实主义特色因为半透明的陈列状态而得以进一步表现：它们的金色与白色外观在暗色木质背景墙的反衬下熠熠发光，并且不时给人悬浮之感。吧台本身由巴塞罗那BD公司设计，其独特形状表现在柱腿的设计，每一根柱腿都是一座雕塑。

  休闲吧的整体主题是轻松、奢华，它的氛围幽暗并带有发光元素。与之搭配，设计师们选择在颜色和外形上有超现实主义特征的作品来装饰（内嵌式的照明灯、人脸形花瓶、各种中空形状的茶几等）。本案设计的一个重要方面是豪华的墙纸、柔和的灯光、金属的帘布和屏风以及家具均能代表特定的区域，如吧台区为乌梅色皮革装饰，休息区为私密的暗色，达利唇形沙发区上方悬吊着深红色的灯，形成一片云的形状。所有这一切都建构了一个内涵一致的空间环境。

Fashion Bar

Designer: Axel Schaefer
Design Company: BERLINRODEO
Location: Berlin, Germany
Area: 140 m²
Photographer: Adrian Schulz

The bar is located in Berlin's Schöneberg district, on the ground floor of a residential and commercial building. Before the reconstruction there was already a run-down bar. While the conversion was done everything new, floor to ceiling, the toilets, the bar-counter, a new entrance, etc.

The owner wanted a bar with the colors red, white and gold. BERLINRODEO then has a glamorous, modern look created. Red wallpaper with baroque patterns are interspersed with gold in the same pattern. This design was also broadcasting the toilets. There were black toilet-objects with gold printing. And the color and material of the walls were taken from the guest rooms. Specially made red hanging lamps have the exact color as the wallpaper. The highlight is the circular color LED light on the ceiling.

The front lounge area is more likely to attend during the day, the rear area is very well attended in the evening and at night.

本案位于柏林的施耐博格区，一幢商住两用建筑的第一层。这里曾经是一家运营不善的酒吧，而现在一切都已翻新：从地板到天花板、洗手间、吧台以及入口。酒吧的业主想要一个红色、白色、金色相间的空间。BERLINRODEO设计公司就为其创建了这个迷人而现代的外观。带有巴洛克图案的红色壁纸上点缀着风格协调的金色图案。这种视觉效果也被延伸至洗手间，其内部物体均为黑色并带有金色印花，墙壁的颜色和材料也与房间一致。特制的红色吊灯与壁纸颜色相同。亮点在于天花板上的彩色LED灯。本案前方的休息区更适合客人白天来此休闲，而后面的空间是傍晚及午夜的好去处。

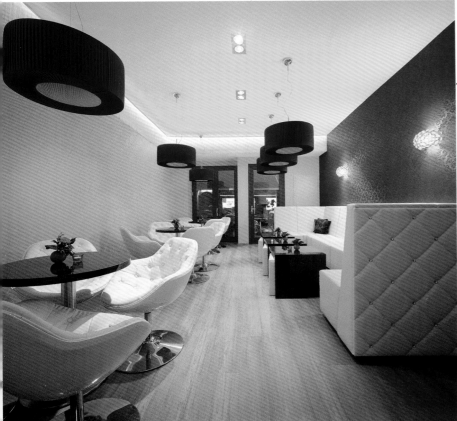

The Mansion

Designers: Rob Wagemans, Joris Angevaare, Erik van Dillen
Design Company: Concrete Architectural Associates
Project Architect: Joris Angevaare
Consultant: E & E projectmanagement
Location: Hobbemastraat 2, the Netherlands
Area: 1300 m^2
Photographer: Concrete Architectural Associates

The Cocktail Bar

One of the Cocktail Bar's most significant features, is the eight wall high mirrors. These mirrors which are "leisurely" leaning against the walls on either side of this space are framed using the synthetic cornice.

The Bar, is like an old-fashioned ornamented cupboard. The synthetic cornice with translucent backing frames the cupboard. This "picture frame" is backlit by LED lighting and provides a gold glowing light feature in this piece of furniture.

In front of this cupboard a lower bar is placed. It is made of chrome steel and in front of it are five tall tables. These tables, with a little square tabletop are part of the bar; they function as pedestals for the presentation of the beautiful cocktails.

The room uses down lights from the ceiling, all placed in a grid to illuminate the tabletops. Other mood lighting is provided by chrome wall lights, with a classic off white lightshade.

All throughout the ground and first floor, the floor is covered by smoked oak wood, in a herringbone pattern.

The Rosé Room

The Rosé Room is quite literally rosé. Draped velour in a skin like pink shade covers the ceiling and walls. This room has a different acoustic feel than other spaces in the building; because of all the fabric in this room, it has a very soft finish.

The cupboard bar again, is one of the main features in this room. The tabletops here are made of Tea Rose Corian; the sides of the tabletops are finished using the synthetic mirror cornice.

Lighting here is provided by down lights in beams across the ceiling.

The Black Bar

The Black Bar is a very chic room, with soft black Napa leather classic chairs and seats. Most of the ceiling is covered by two huge images. The image is a modern interpretation of the ceiling in the Sistine chapel. The backlit ceiling is made of a special synthetic material that is used for light cases. Here the glowing ceiling makes the whole space feel very comfortable.

The image used on the ceiling is also used for the carpet. Walls are covered with silver lily patterned wallpaper that is painted black.

To mask the harshness of the outside world, the windows are foiled with a dark grey vinyl and also covered by black lace curtains.

section AA

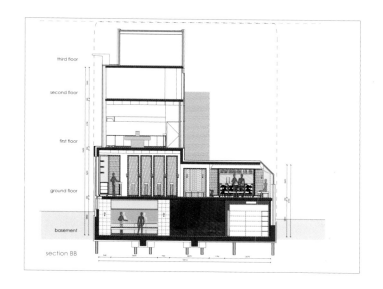
section BB

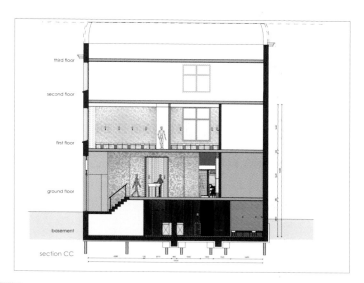
section CC

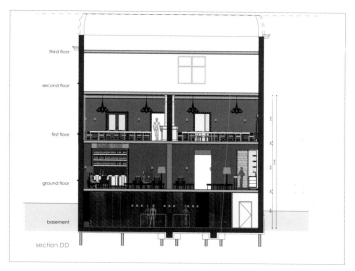
section DD

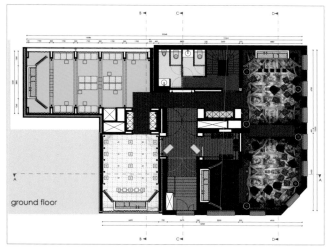
ground floor

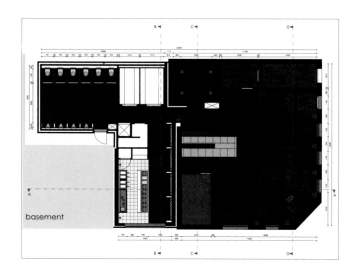
basement

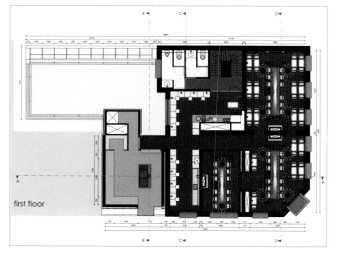
first floor

鸡尾酒酒吧

鸡尾酒酒吧最显著的特征之一是它八面如墙高的镜子。这些镜子随意地依靠在空间四周的墙壁上，边框由合成的檐板镶嵌制成。

这间酒吧如同一款旧式的装饰橱柜。合成檐板与透明背板构成了橱柜的边框。这种"相框"由LED灯光照亮，在这款家具上映射出金色的光亮。

在橱柜的前面设置有一个较低的吧台。它是铬钢材质的，在其前面摆放着五个高桌，并配有一个小型的方形桌面，它们作为酒吧的一部分为展示精美的鸡尾酒提供了基座。

这间房间采用射灯照明，所有射灯被摆放在一个格子内，照亮桌面。其他的气氛光源来自铬合金墙灯，营造出一种经典的纯白光影。

整个底层和一楼地面皆由栎木铺置而成，呈人字形图案。

玫瑰厅

玫瑰厅是一间真正意义上的玫瑰之屋。天花板及墙面垂挂着粉色的丝绒。同本建筑内其他空间相比，这里有着一种不同的声音效果；因为其内部质地非常柔软。

橱柜吧台也是这间房间的主要特色之一。在这里，桌面采用淡橙红色的可丽耐材质；桌边由合成镜面檐板装饰。

玫瑰厅的照明是由天花板上的束状灯光提供的。

黑色酒吧

黑色酒吧是一个非常时尚的空间，有柔软的黑色光面皮革制成的经典坐椅。大部分天花板被两个巨大的图像覆盖。这个图像是对西斯廷教堂的天花板的现代诠释。背光天花板由一款特殊的合成材质制成，使得整个空间充满了舒适感。

地毯也采用天花板上的图案。墙面覆盖着银色百合图案的墙纸，被绘制成了黑色。

为了隔绝外部世界的纷扰，窗上装饰着深灰色的乙烯基材质，并配有黑色蕾丝窗帘。

W Lounge Bar

Design Company: Concrete Architectural Associates
Location: W Hotel, Leicester Square, London, UK
Project Team: Rob Wagemans, Jeroen Vester, Ulrike Lehner, Erik van Dillen, Melanie Knüwer, Jari van Lieshout, Sonja Wirl, Nina Schweitzer
Lighting Consultant: Maurice Brill Lighting Design
Photographer: Ewout Huiber

This is not so much about the bar; it is the 37 meters chesterfield couch that defines the social landscape for guests and friends to live, meet, and mingle. Everything, including the end-grain oak flooring and gold leaf ceiling, naturally follows the winding landscape. Loose elements cut away from the chesterfield landscape, function as coffee tables and offer a platform to the cylindrical fireplaces that follow the shape of the columns surrounding the couch. The chesterfield defines an important part of British heritage and the way the columns run through the chesterfield suggests the sofa was there long before the modern building was. The lounge bar is a good place to stay at any hour of the day. A striking framework of vertical blinds can be set in different positions and creates different light scenes according to the time of the day and use of the space. A third, loose fragment cut away from the chesterfield landscape has become a golden lit bar. It transforms from a breakfast bar in the morning into a bar at which one can prepare for a wild night out.

　　W酒吧内的社交场景由一个37米长的切斯特菲尔德大沙发勾勒而出，宾客和朋友们可来此享受生活、聚会见面、交往接触。这里的一切，包括端头木纹的橡木地板和金叶天花板，都自然地符合着蜿蜒的景致设计风格。一些分散的元素独立于大沙发场景之外，它们被设计成咖啡桌，并为圆柱形的壁炉提供了底座，这些壁炉与环绕在沙发周围的柱子形状相吻合。切斯特菲尔德市是大不列颠历史的重要组成元素，酒吧里以柱子环绕于沙发周围的方式表明：早在这座现代建筑出现之前，切斯特菲尔德大沙发就存在已久了。一天之中无论何时，W酒吧都是一个好去处。直立式百叶窗非常醒目，可以任意调整其位置，根据不同时间和使用场合营造出不同的光线场景。另一个独立于大沙发场景之外的元素被设计成金色的光吧。清晨的早餐吧在夜晚会变成一个任你彻夜狂欢之地。

WYLD Bar

Design Company: Concrete Architectural Associates
Location: W Hotel, Leicester Square, London, UK
Project Team: Rob Wagemans, Jeroen Vester, Ulrike Lehner, Erik van Dillen, Melanie Knüwer, Jari van Lieshout, Sonja Wirl, Nina Schweitzer
Lighting Consultant: Maurice Brill Lighting Design
Photographer: Ewout Huibers

When the daily drag is done, guests can dance the night away at WYLD bar. The bar and two-storey high liquor cabinet, with jewel boxes to highlight the selection of drinks, overlook great Leicester Square — heart of the London cinema land. WYLD's interior look combines the red carpet feel with spicy red & black leather furnishings and a grand finale diameter of 3 meters disco ball. A circular booth surrounds bespoke cocktail tables that light up in red. You can stock away glasses and ice inside its stainless steel pockets. The high level walls are covered with black sequins referring to the West End and moving according to the beat of the music. The DJ looks down onto the dancing crowd and when the music starts to play, the sequins wall opens up to create a stage. Adding to the intimacy of the space, fiber optic strings hang from the windows and create a cocoon.

当人们结束了一天的忙碌后,可来WYLD酒吧跳舞消遣长夜。酒吧里两层高的酒柜配以珠宝匣式的装饰来突出饮品多样的选择,俯视着位于伦敦剧院之地中心的莱斯特广场。WYLD的内部装修结合了红毯盛会般的隆重感,使用红黑色真皮家具和一个硕大的直径为3米的迪斯科球。被红色灯光照亮的定制的鸡尾酒桌四周被圆台环绕。客人可以在圆台的不锈钢凹洞里放置玻璃杯和冰。高层墙壁上挂满了象征伦敦西区的黑色亮片,并随着音乐节拍而舞动。音乐主持人俯视着跳舞的人群,当音乐响起来时,亮片墙打开,舞台出现。 窗上一行行的光纤线为酒吧勾勒出茧的形状,使空间亲近感倍增。

Republic Gastropub

Design Company: Elliott + Associates Architects
Location: Oklahoma City, Oklahoma, USA
Area: 581.85 m² (497.68 m² ground floor + 84.17 m² mezzanine)
Photographer: Scott McDonald

Architectural Concept:
1. The approach to the design supports the proposed name, menu, and spirit of the project concept.
2. The design will be the "coolest sports bar in America".
3. The concept is developed around the "spirit of sport."

Design Feature:
1. Power and grace.
2. Strong muscular qualities.
3. A place with spectacle where you can cheer your team on with all your friends.
4. A place for a comfortable, memorable dinner with signature food and beverage.
5. It will be a great place to hang out.

Design Details:
1. Central focus is 200" projection screen and 103" plasmas with 42" eye level LCDs for bar patrons.
2. 2-storey beer cooler and beer bottle display.
3. Bar for 30 patrons.
4. Bubble Wall.
5. Seating for 171 — 88 in booths and 53 at tables and chairs, seating for 16 on the outdoor patio.
6. Patio as an outdoor room with 4 plasma TVs and fans built-in to the TV enclosure. 2.44 meters Nellie Stevens hedge provides a sun and wind break and makes the patio more of a surprise for patrons.
7. Booths are paperstone with coarse leather seats and white oak wood tables.
8. Floor is dark grey marble terrazzo using a dark grey matrix.
9. Ceiling is painted exposed steel structure and metal deck with suspended sound panels.
10. Bar is copper, top and face, with a steel plate footrest and soft downlight from under bar.
11. South accent wall is the bubble wall.
12. Separation walls at bar and booths are paperstone.
13. Kitchen wall is paperstone.
14. Multi-color LED lighting in corridor to toilets.

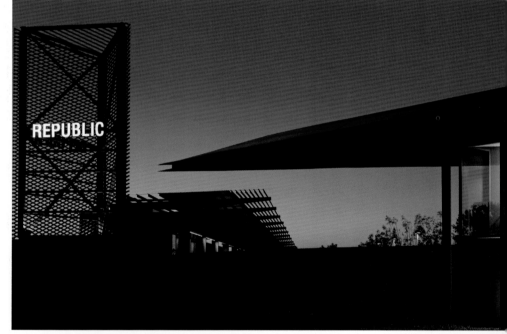

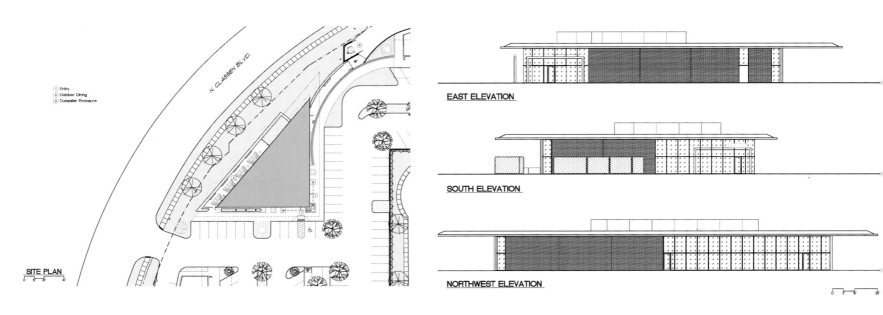

SITE PLAN
1. Entry
2. Outdoor Dining
3. Dumpster Enclosure

EAST ELEVATION

SOUTH ELEVATION

NORTHWEST ELEVATION

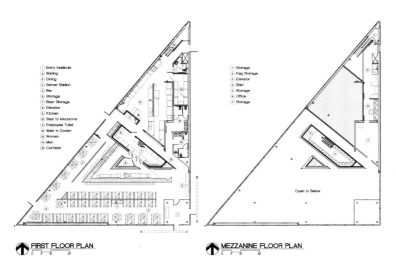

FIRST FLOOR PLAN   MEZZANINE FLOOR PLAN

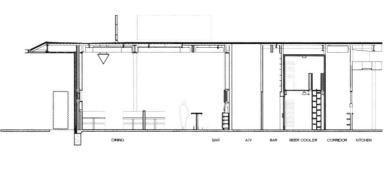

SECTION LOOKING WEST

设计理念:
1. 设计方案诠释了店名、菜单和项目设计理念。
2. 本案将成为"美国最著名的体育酒吧"。
3. 全部设计围绕"体育精神"的理念展开。

设计特色:
1. 体现力量与风度。
2. 强烈的阳刚之感。
3. 可以观看比赛,与朋友一起为心中的队伍助威。
4. 在此用餐,空间舒适;招牌菜品与饮品使人印象深刻。
5. 总之,这里是你的娱乐首选之地。

设计细节:
1. "镇店法宝"是200"投射大屏和103"等离子电视以及42"液晶电视。
2. 双层啤酒冷却机和啤酒瓶展示架。
3. 可容纳30位常客的吧间。
4. 气泡墙。
5. 全部客容量为171人,包间88人、普通坐椅53人、室外平台16人。
6. 室外平台备有4台等离子电视和固定风扇;2.44米高的围墙可防风遮阳,给人带来更多的惊喜。
7. 包间台面为再生纸材料,坐椅用粗粒面皮革包覆,餐桌为白栎木质。
8. 地面为黑灰色大理石和杂石水磨地面。
9. 天花板使用涂漆外露钢架结构和有悬挂式吸音板的金属承板。
10. 吧间从天花板到内部墙面,多处采用铜料装饰,并配有钢制脚凳,吧间下部有柔光照明。
11. 南墙是与其他墙体迥异的气泡墙。
12. 吧间和包间的隔断墙为再生纸材料。
13. 厨房墙使用再生纸材料。
14. 从走廊到盥洗室为复色LED照明。

Bar Fou Fou

Designers: Peter Ippolito, Gunter Fleitz, Tim Lessmann, Hakan Sakarya, Yuan Peng
Design Company: Ippolito Fleitz Group—Identity Architects
Location: Leonhardstr. 13, Stuttgart, Germany
Area: 102 m²
Photographer: Zooey Braun, zooey@zooeybraun.de

The Fou Fou is a champagne bar situated at the very edge of Stuttgart's red light district. Opening onto a small square, it occupies a corner building that formerly housed an antique shop. The interior design was inspired by the building's location and history, resulting in a space where boudoir meets salon. The Fou Fou makes optimum use of the limited space at its disposal: four separate salons, each with its own unique character, stretch over three storeys.

You enter the Fou Fou straight into an L-shaped barroom arranged around a large bar counter. Mirrors affixed to the wall behind the bar expand the space further still. The entrance area is heralded by a large, fabric-covered ceiling light that serves as a room opener. The bar itself is upholstered in a metallic greenish-gold diamond pattern, which is continued on the ceiling in the rear area of the barroom. The walls are executed in a complimentary shade of classic green. This forms an effective backdrop for the white, original case tree windows. The bar can be opened up to the outside during the summer, enabling customers to be served outside. A continuous bench runs along the outer wall of the bar area, enabling maximum use of the limited space. At the rear of the room is a newly constructed, enclosed stairwell. A black, sculptural element serves as a hinge directing one flight of stairs to the floor above and another down to the floor below.

The FouFou is a space that successfully combines genteel glamour with modern appeal. The different lounge areas boast ample room to savour a good glass of champagne in a suitably sophisticated setting, as well as serving the needs of club-goers on the weekend.

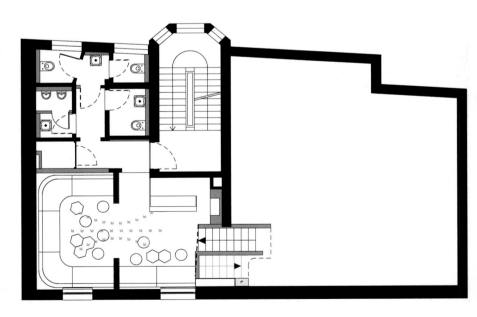

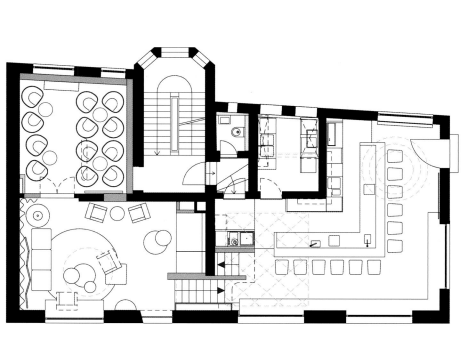

Fou Fou吧是一家香槟酒吧，紧邻斯图加特的红灯区。该酒吧朝向一个小广场，位于拐角处的一座楼内，以前在这座楼内曾开有一家古董店。本案内部设计的灵感来自于该建筑本身的地理位置和历史，既包含化妆间又设有沙龙，既最大限度地利用了有限空间又做到了自由设计：它的四个独立沙龙占据了三个楼层，每个沙龙都独具特色。

进入Fou Fou，即进入了一个L形吧间，整个吧间沿一个大型吧台而建。吧台后面墙壁上的镜饰使空间增加了纵深感。入门玄关处上方有大型布艺顶灯与其呼应并作为入口装饰。吧间本身由具有金属质感的金绿色菱形纹装饰，这些饰样一直延伸到吧间后部的天花板。墙面则是古典的绿色调，这样的颜色设计为吧间新颖的白色格窗提供了生动的背景。在夏季，吧间可在室外营业，即顾客可在室外享受服务。吧台区域沿外墙有一体的长椅，最大限度地利用了有限的空间。吧间后部是一个新建的封闭式竖直楼梯，楼梯的黑色雕塑装饰一端连接上层楼梯，另一端连接下层楼梯。

Fou Fou酒吧成功地融合了绅士气质与现代魅力。各个休闲区域拥有宽阔的空间，使顾客在各种高雅的气氛中尽享香槟带来的美妙感受，同时也能满足吧迷们周末休闲的需求。

 BAIXA

Designer: José Carlos Marques Cruz
Location: Porto, Portugal
Area: 120 m²
Photographer: FG+SG – Fotografia de arquitectura

For years, Porto Downtown was a day-time zone. This was the civic center and buildings were ocuppied by offices and commercial stores. The gentrification phenomenon made people move to cheaper zones, leaving empty buildings and desert atmosphere at night.

With the advent of new rehabilitation politics and a series of incentives, new kinds of businesses were created and two years' time turned the whole civic-center into the bohemian-center.

BAIXA stands for "Downtown" and its inserted in a context where buildings have a Paris beaux-arts style.

The idea for the project was set on that Parisian ambience as a starting point.

The space was divided in two rooms with the facilities between them. One is destinated to relax and have a drink. The other one is to dance.

The first room, next to the entrance works as an extension of the exterior environment. Decorated with plaster frames on the walls and rosettes on the ceilling, the Deco Style is abruptly interrupted by a sculpted tunnel that links into the other room.

With a whole different environment, the second room is where the dance floor is located and was inspired on a Sol Lewitt installation. The mirror ceiling gives it the "Disco look" that was intended.

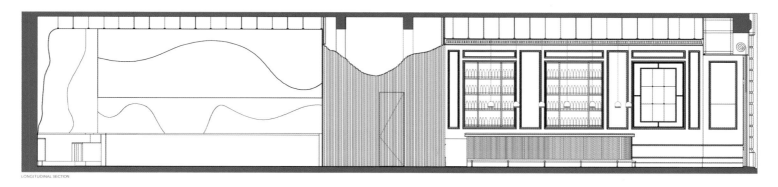

LONGITUDINAL SECTION

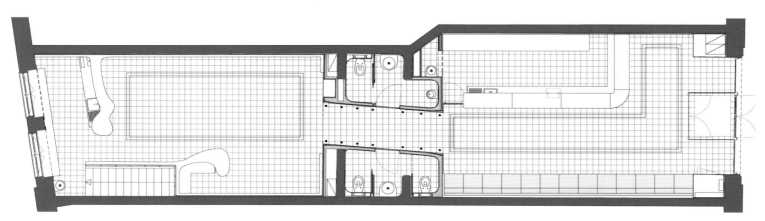

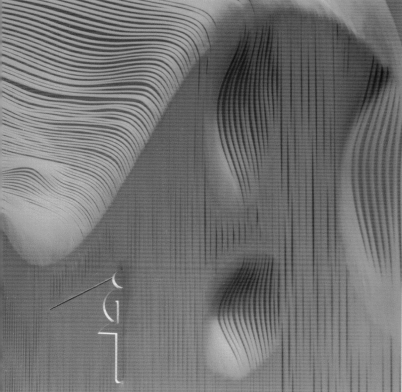
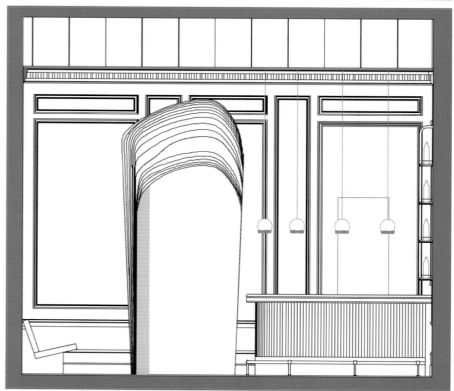

TRANSVERSAL SECTION

多年来，波尔图中心都是一个日间活动区域。作为市中心，各栋大楼满是写字间和商店。"中产阶级化"现象出现后，人们迁往周边地价更加便宜的地区，使得该地住宅建筑人去楼空，到了夜间更是一片萧肃。

随着该地区政府复兴计划的出台和一系列刺激发展措施的施行，各种新型商业次第出现，整个市中心就又成了小资生活的聚居地。

BAIXA意为"市中心"，它地处鳞次栉比的楼宇之中，处处体现着时尚不羁的巴黎风格。

本案的设计理念即以打造一种巴黎风格为出发点。

整个室内空间分为两室，功能区位于二者之间。一室用作休闲餐饮区，另一室为舞厅。

休闲餐厅区在主入口旁边，是室内环境的延伸。墙体以石膏模型装饰，天花板则是玫瑰造型，但这样的造型空间在以雕像装饰的通道处戛然而止，而后者通向另一个空间。

舞厅的风格截然不同。其设计灵感来自于索尔·勒维特的概念艺术。天花板上的镜面装饰更烘托出空间的气氛，也令设计者的初衷尽显无遗。

MUGEN

Designer: Yusaku Kaneshiro
Design Company: Yusaku Kaneshiro+Zokei-syudan Co.,Ltd
Location: Ebisu, Tokyo, Japan
Area: 58 m²
Photographer: Masahiro Ishibashi

A big dignified Bubinga wood table has been planted in the room where mirror effects are interwoven.
Lights from creative arts and chandeliers reflects diffusely in the ebony space.
These prevailing black color and dotted lights impress an image of infinitely far beyond.

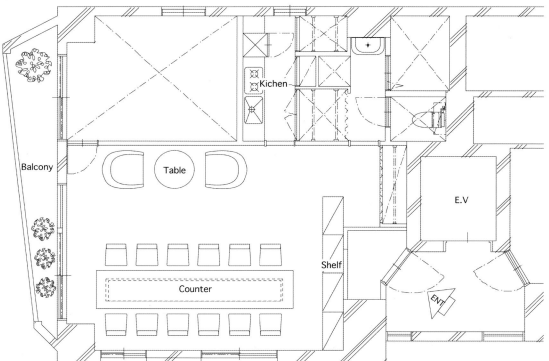

一张彰显气派的非洲花梨木长桌被置于镜面交错的房间。

创意灯饰让这个深色空间充满神秘色彩。

这种经典的流行黑与装饰灯融合在一起创造出一种无限延伸的视觉效果。

Twenty Five Lusk

Designers: Cass Calder Smith, Bryan Southwick, Barbara Turpin-Vickroy
Design Companies: CCS Architecture, 44 McLea Court
Location: San Francisco, USA
Photographer: Paul Dyer

On Lusk Alley in San Francisco's South of Market district, a 1917 smokehouse and meat-processing facility has been renovated to become Twenty Five Lusk. The 265-seat new American restaurant and bar is an unexpected gem in the urban fabric. CCS Architecture crafted the two-level space, weaving graceful forms and sophisticated materials through the massive, historic, warehouse structure. The interior emphasizes a counterpoint between the new palette of polished stainless steel, glass, white plaster, leather, mirror, faux fur, and slate and the existing elements of brick, concrete and rough-sawn timber.

The architects created a large, glass entrance, cutting into the existing building exterior; the canopy bends up at its leading edge to become the restaurant's signage. Windows were enlarged and added along the facade to animate the interior with natural light and allow views. Inside, a large wedge from the upper floor makes an open connection between the lower level lounge and the dining room upstairs. Entering the space, guests take in simultaneous views of both.

The architecture sets up a notable contrast between the dramatic vertical space and the single-height areas, allowing guests to experience the space in its totality while providing intimate spaces to explore. CCS transformed the entire 1393.55 square meters warehouse to accommodate new uses. Twenty Five Lusk occupies the first and second floors, and the third level has been designed as 483.10 square meters of creative office space.

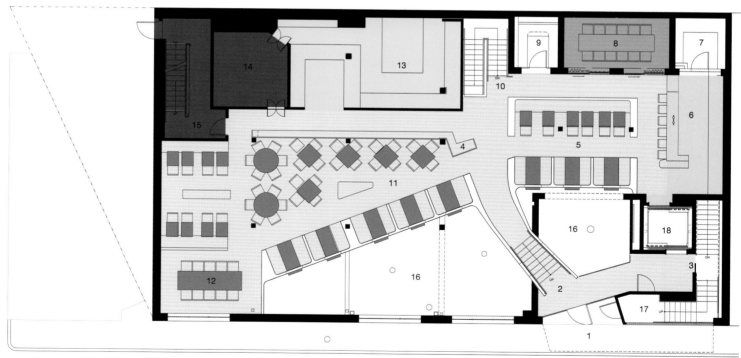

FIRST FLOOR PLAN

LUSK ALLEY

1. MAIN ENTRY
2. STAIRS TO MAIN DINING
3. BACK STAIRS TO LOUNGE
4. HOST
5. BAR SEATING
6. BAR
7. LIQUOR STORAGE
8. PRIVATE DINING
9. WINE ROOM
10. MAIN STAIRS TO LOUNGE
11. MAIN DINING
12. COMMUNAL TABLE
13. GLASS-ENCLOSED KITCHEN
14. WARE WASH
15. EXIT STAIRS
16. OPEN TO LOUNGE BELOW
17. STAIRS TO OFFICE
18. ELEVATOR

- main dining + entry
- back of house
- kitchen
- bar
- private dining

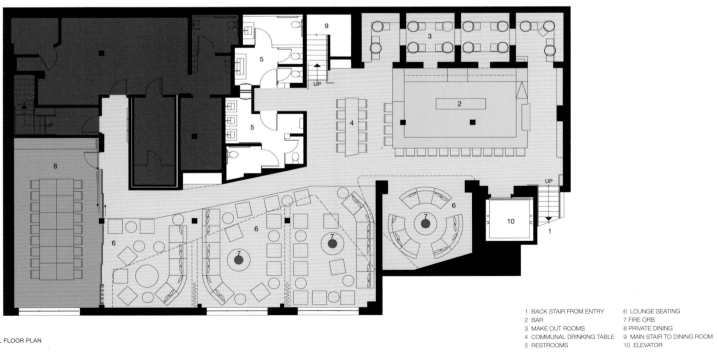

LOWER LEVEL FLOOR PLAN

1 BACK STAIR FROM ENTRY
2 BAR
3 MAKE OUT ROOMS
4 COMMUNAL DRINKING TABLE
5 RESTROOMS
6 LOUNGE SEATING
7 FIRE ORB
8 PRIVATE DINING
9 MAIN STAIR TO DINING ROOM
10 ELEVATOR

lounge
office/back of house
bar
private dining

在旧金山的拉斯克巷南商业区，一座名为25拉斯克的餐厅和酒吧由1917年建成的熏制室和一些肉类加工设备改造而成。这间容有265个坐席的全新美式餐厅和酒吧是都市中的一颗明珠。CCS建筑公司建造了双层空间，在大规模的历史性仓库构建中采用曲线优美的框架和先进的材料。内部通过抛光不锈钢、玻璃、皮革、白石膏、镜子、人造毛皮、板岩与已有的砖、混凝土和棱角粗糙的木材形成对比。

建筑师打造出的大型玻璃入口，融入现存的建筑外立面中；弯曲的顶篷成为餐厅的招牌。窗户被加大的同时数量也随之增加，利用自然光使室内灵动起来并扩充了人们的视野。在内部，一个大型楔形结构从上层延伸出来，将较低位置的休闲吧和楼上的空间连接起来。当客人步入时会自然而然地留意到两者的存在。

此建筑体现了纵向空间和单高度区域的鲜明对比，客人于此体验空间的同时也保留了个人的私密空间。CCS建筑公司将1393.55平方米的仓库改装后使之适应新的用途，25拉斯克餐厅占据了一层和二层，第三层被设计成占地483.10平方米的创意办公空间。

Disko Bar Mladost

Designer: Djordje Gec
Design Company: Fluid:Architecture
Location: Belgrade, Serbia
Area: 120 m²
Photographer: Ana Kostić

Disco bar Mladost is located in Karadjordjeva Street, in the ramshackle neighbourhood of Savamala. The project was immersed in the environment with an idea to build on the existing spatial image and to liven up the world of city's dark bars.

The task was to transform the existing space into a disco bar with a clear identity, without extensive construction work. The space consists of two units, "bar" and "basement" whose relationship is suggested by central spatial elements, a retro bar and a DJ booth, and then continued in the stairway used by the guests to descend into the "basement" and go from one unit to another. Therefore, two units are connected, but at the same time there is a possibility of separate activities.

Traces of formwork are visible on the bar, along with concrete surfaces, exposed brick and steel reinforcement which are dominant within the space. Light comes into the room directly through the ground glass bar and through the reinforced mesh placed under the ceiling. Raw image of the space is achieved by combining the material and light effects.

　　Mladost迪斯科酒吧位于Karadjordjeva街道，靠近Savamala。该项目利用现有的空间建造，使其恰当地融于周围环境，亦使其在该市偏昏暗的酒吧中显得更有生气。

　　该项目须在现有的空间中建造一个迪斯科酒吧，并使其个性突出，但是建筑规模不能太大。空间由酒吧和地下室两部分组成。受中央空间元素的启发，复古的酒吧和音乐主持人台由一条通道连接。顾客可由该通道从酒吧进入地下室，在这两个空间中自由穿梭。因此，两个空间被联系到一起，但同时又存在独立举办活动的可能。

　　在空间中可以看到模板的痕迹，伴随混凝土表面、暴露的砖块和钢筋主导整个空间。酒吧的磨砂玻璃以及天花板下的加固网板使室内灯火通明。建筑材料和灯光的巧妙结合，给顾客与众不同的感觉。

INDEX

3GATTI

Francesco Gatti was born in Rome in 1973. He graduated with honors degree in ROMA TRE university. In 2002 he established 3GATTI. In 2004 he opened a new branch of the office in Shanghai where he completed many projects as the "Red Wall", the "Shenyang Shopping Mall", the "In Factory" old factories renovations, the "KIC Plaza" park, and prestigious interiors as the "Red Object" space, the "Hightex" textile group concept shop, Hangzhou "The Cut" interactive disco club, "ZeBar" and "Alter" fashion store.

GATTI won many competition and awards as: the "Buchanan Underground Station" (Glasgow, Scotland), the "Tra la murgia e il mare" urban development (Andria, Italy), the "500M³ design" GBD art residential district (Beijing, China), the "06 Modern Decoration Interior Design Media Prize" (Shenzhen, China) and recently won the international competition for the construction of the Automobile Museum in Nanjing. His work was published in many international reviews.

Today GATTI teach in ROMA TRE University (Italy) and in Tongji and Jiatong University (China) where he also participated as master of the Archiprix International.

Antonio Di Oronzo

Antonio Di Oronzo came to New York from Rome (Italy) in 1997 and has been practicing architecture and interior design for eighteen years. He is a Doctor of Architecture from the University of Rome "La Sapienza", and has a Master's in Urban Planning from City College of New York. He also holds a post-graduate degree in Construction Management from the Italian Army Academy.

In 2004, Antonio founded the award-winning firm bluarch architecture + interiors + urban planning, a practice dedicated to design innovation and technical excellence providing complete services in master planning, architecture and interior design.

At bluarch, architecture is design of the space that shelters passion and creativity. It is a formal and logical endeavor that addresses layered human needs. It is a narrative of complex systems which offer beauty and efficiency through tension and decoration.

3SIX0 Architecture & Design

3SIX0 is an architecture firm headed by Kyna Leski and Chris Bardt. Our work is best defined by the spatial idea that is created out of the forces, content and limits unique to each project. This spatial idea acts as a guide to the many decisions through all stages — from sketch through construction. In this way, one does not perceive an applied theme or style. Instead, one experiences a sense of "place" that is integral to the situation, directly through the workings of the architecture: through phenomena — light, sound, view, shadow, balance — and through images triggered by the imagination.

Archinexus

Archinexus, founded in 2004 by Mr. Tai-Lai Kan, a leading-edge design practice based in Taipei,China, is a creative and collaborative design firm with extensive practical experience, especially renowned for strong design across a variety of high-profile projects. Dedicated to innovative problem solving and design excellence, Archinexus is a team of highly qualified design professionals daring to dream as well as those who are capable of making dreams come true.

Andy Martin Architects

The AMA design philosophy embraces creative intelligence. AMA's unique design identity is demonstrated through a wide range of UK based and international projects that offer brand strategic, creative, market appropriate solutions. Our insight and breadth of knowledge extends to all facades of architecture and design.

Andy Martin Architects is responsible for the creation and evolution of some of London's popular restaurants including ISOLA, QUO VADIS and MASH. AMA works in collaboration with London's most renowned entrepreneurs and chefs, Oliver Peyton, Chris Corbin, Jeremy King and the Hart Brothers. The practice has further completed numerous influential projects internationally.
www.andymartinstudio.com

Architect Javier Serrano Orozco

Born on March 22nd 1982, Mexico City.
Bachelor Degree in Architecture, Universidad Iberoamericana Mexico City (2001-2006).
Established CHEREMSERRANO arquitectos in 2004 with partner Abraham Cherem Cherem.

Archi-Tectonics

Winka Dubbeldam is the principal of Archi-Tectonics NY, founded in 1994 and Archi-Tectonics NL, founded in 1997. Dubbeldam is a graduate of the Academy of Architecture in Rotterdam (1990), and received a Degree in Master of Science in Advanced Architectural Design from Columbia University, NYC in 1992. She has lectured extensively and taught at the Masters Programs of Columbia University, NYC and Harvard University, Cambridge and currently holds the position of Director of the PP@PD, the Post-Professional Program at the University of Pennsylvania, Philadelphia. She has also served as juror in design competitions among which the AIA and the Architecture League, NY, as well as in a multiplicity of reviews at International Architecture Schools. Winka is also an external examiner for the RIBA/ARB at the Architectural Association, London, and serves on the Board of Directors of the Institute for Urban Design, NYC.

Axel Schaefer

Axel Schaefer, interior and furniture designer, is much more than a local hero of the Berlin design scene. His company BERLINRODEO, specializing in integral architecture and design, is one of the most innovative spots for modern architecture and groundbreaking interior decoration.
Axel Schaefer, who studied architecture in Berlin, has accumulated experience with well known, internationally operating design and architecture studios. He founded his company BERLINRODEO in 2006. All work is focused on the requests of his clients, integral design and innovation.
Integral design – Axel Schaefer's definition of integral design is an intense debate with the client, wherein he takes into consideration their individual style, wishes, expectations and ideas. Furthermore, it is the exact handling of the space and the specific knowledge about the local conditions. With knowledge of the client's needs and an understanding of the requirements of rooms, one has the ability to deliver an individual, tailored, and collaborative result.

B3 Designers

B3 Designers creates spaces derived from a marriage of aesthetics and functionality, tailored to the client's needs. Our architectural interiors are spatial solutions to the demands of each project's use, projecting an image that fits with the company's brand values.
Established in 2002, the company specializes in unique and contemporary designs for hotels, restaurants, bars, retail spaces and exhibitions. Experience within these fields has led us to work with diverse clients, ranging from those that have a clear idea of what they want and simply need us to facilitate it, to those who would like us to develop the brand alongside the interior.

BCV Architects

Baldauf Catton von Eckartsberg Architects is a San Francisco based design firm known for the diversity of scales at which it works — from the master planning of large urban projects to the tableware used in our restaurant designs. The firm's principals pursue this range of work because they believe architecture and design are richest when they are informed by the breadth and complexity of human experience. An interest in the broad approach to a design problem lends itself to the multi-disciplinary character of BCV Architects and to experiences in urban design and planning, architecture, interiors, furnishing, and graphic design.

Bells & Whistles

Established in 2000, Bells & Whistles is a leading San Diego based design/build studio that is dedicated to creating quality, hand-made, modern interiors. Working primarily with commercial spaces, the firm's projects have included bars, restaurants, boutiques, entertainment venues and salons with the occasional residence thrown in the mix.
Bells & Whistles is the design trio of Jason Lane, Barbara Rourke, and Jason St. John. All are seasoned designers, artists, and craftspeople whose capabilities range from wood and metal working to lighting, furniture and environmental design to sculpture and art.
Bells & Whistles is capable of taking on projects of any scale from concept to completion. The designers' goal is to create amazing interiors that showcase their talents and vision while successfully realizing their clients' dreams.

CCS Architecture

CCS Architecture is dedicated to excellence in architecture and design. Since its inception in 1990, CCS has designed a diverse range of public and private buildings and interiors for LIVING, WORKING, EATING, and MORE. The firm has gained international acclaim for the architectural and commercial success of its restaurant projects, while the uniqueness of residential, commercial, and mixed-use projects has met with an unusual degree of owner satisfaction and media praise.
CCS seeks to explore opportunities of maximum potential and express them at a scale appropriate to each project. The work is firmly based in the modernist idiom, where innovation and creativity are balanced by common sense and experience. The firm is known for creating projects with exceptional spatial and material qualities, and for providing outstanding, professional service.

Concrete Architectural Associates

Concrete Architectural Associates is founded in 1997.
The present director of Concrete Architectural Associates, Rob Wagemans, was born in Eindhoven on 13th of February 1973, he is a Master of Architecture Utrecht.
Erik van Dillen (at this moment he is just creatively involved with Concrete), interior architect, was born in de Bilt on 27th of April 1960, catering industry skills in the kitchen, painting restorer.
Concrete originally was founded by Rob Wagemans, Gilian Schrofer and Erik van Dillen. They met each other by a not realized project, a head office in Amsterdam for Cirque du Soleil. Gilian Schrofer left concrete in 2004 to start his own company.
Concrete Reinforced is founded in 2006 by Rob Wagemans and Erikjan Vermeulen (present co-director of Concrete Reinforced).Erikjan Vermeulen is a Master of Architecture. He worked for different architects to start his own company in 2003.

 ### Cor & Partners

Cor is a company that works to "make things possible": we want to be able to provide creative and optimal solutions to the opportunities facing us, with the objective of getting them into reality.
Cor is a team of creative and skilled, able to respond to multiple problems. We arecommitted to working in an interdisciplinary way in order to solve the many variables that a project entails.
Cor is a consultancy that works seamlessly in the world of Architecture, Planning and Engineering and in Business Management and Audit. To do so, rethink and brand updates, verifying and managing their identity and position in the Web.

 ### crossboundaries architects

crossboundaries architects is a young, Berlin and Beijing based team of international architects aimed at linking the professional and design experience of trained architects and consultants to the challenging possibilities of the prospering Chinese economy as well as to millennia of Chinese cultural heritage and the lively, multifaceted Chinese culture. Our goal is to develop high-quality solutions, derived from Chinese culture and architectural needs, based on western know-how and experience.
While continuing to be active in the area of architectural design in China, crossboundaries architects have begun to vigorously explore and participate in culture-exchanging research, programs and art exhibitions, in order to enrich its practice. As the name indicates, crossboundaries architects stay in an elaborated contemporary definition of architectural practice with a broad view that looks beyond.

 ### designLSM

designLSM is a multi-disciplined design practice comprising interior designers, architects and graphic designers — providing the highest quality integrated and innovative design solutions. We are passionate about what we do and have consistently delivered high quality projects for restaurants, hotels, bars, retail and offices for over two decades. designLSM work throughout the UK and internationally.

 ### Dexter Moren Associates

Dexter Moren Associates is a London based design practice offering urban design, architecture and interior design. Since its establishment in 1992 the practice has completed a wide range of projects including offices, shopping centres, mixed tenure housing, urban renewal developments and significantly has developed a reputation as one of London's leading hotel & leisure architects.
As the UK member of Perspective DMA is able to offer a seamless service across borders via its network of offices throughout Europe. Independent of Perspective the practice is currently designing a range of luxury hotels & resorts in the Middle East, North Africa and Asia.
The practice aims to deliver intelligent, innovative and sustainable solutions whilst ensuring a collaborative and responsive approach tailored to the aspirations of its clients. The team is headed up by four directors and four associates who bring collective skills in design, sustainability, marketing & finance. DMA is a Chartered RIBA Practice and is ISO 9001:2008 accredited.

 ### Dimitris Naoumis

Dimitris Naoumis was born in Athens, Greece.
As an industrial designer, he has worked on the creation of a variety of objects which have been produced and sold in Europe, America and Saudi Arabia.
The last years he works in his Architect Studies Office in Athens, which is responsible for the design of many bars, restaurants, offices, shops, houses etc.
Through the analytic research of the main characteristics of each project, the Architect Studies Office, forms a proper design study and a realistic presentation which shows a full view of the completed project, even before the beginning of the construction.
The supervision of the project and the effective co-operating are factors that reserve the quick and high quality constructing rhythms, using of the best quality materials for each project.
The combination of elements that ensure the aesthetic unity and the functionality of each interior space is very important during the design process. The new materials and the innovative technologies create a style that is contemporary and also diachronic.
The office creates conceptual spaces with scenographic environments.

Dyer-Smith & Frey

Quality, great attention to detail and creation of emotions are the main focus of Dyer-Smith & Frey. The two designers look to capture elegance within the details of a product and thereby tell an inspirational story.
James Dyer-Smith and Gian Frey create room concepts that radiate both emotion and strength. They design unique furniture, produced in a range of price segments from serial production to individual hand-made pieces. Realisation of Corporate Brandings and tailor made solutions, from logo design right through to the website and other visualisations, are further speciality.
James Dyer-Smith and Gian Frey obtained their degrees in Product and Industrial Design at the renowned Zurich Hochschule of art.
James Dyer-Smith gained his post-studies work experience at MACH Architecture in Zurich and earlier at the company of Tyler Brulé in London.
Gian Frey gained his work experience at Christophe Marchand Product Development in Zurich, at the Architectural Digest Magazine (AD) in Munich. Post-studies he was part of Matteo Thun & Partners team in Milan.

 ## Elliott + Associates Architects

Established in 1976, Elliott + Associates Architects is a full-service architectural firm of licensed architects, interior and graphic designers and support personnel. The firm has designed award-winning projects for corporate clients, various arts organizations, museums, and public spaces.
Elliott + Associates Architects creates special environments - architectural portraits - revealed as expressions of the client. Elliott + Associates key members are Rand Elliott, FAIA, who is Principal-in-Charge and Design Architect and Bill Yen, AIA, Senior Associate. Mr. Elliott has been principal of his own firm for 35 years. The firm's projects have won 258 international, national, and local awards including ten National AIA Honor Awards.

 ## Fluid:Architecture

Fluid:Architecture is a Belgrade based interdisciplinary architectural design studio.
Our approach is to explore the many facets of every project and its specificity and to search for original solutions, no matter the size and the type of the project. Unique character of each project becomes the starting point for an architectural idea which forms the basis for visual aesthetic. Identifiable style and logic of our projects are clearly seen. Our main priority is ability to communicate with every client and to understand their needs in order to find unique and functional design solutions.
www.fluid-architecture.com

 ## FGMF

Created in 1999 by fellow students from FAU-USP, Forte, Gimenes & Marcondes Ferraz (FGMF) produces a contemporary architecture, without any restraints regarding the use of material and building techniques, seeking to explore the connection between architecture, environment and mankind. During these years we've received significant awards, among which many from the Instituto de Arquitetos do Brasil (IAB), Living Steel (IISI), Editora Abril and ABCEM. Recently, FGMF has been chosen as the only Brazilian office to integrate the distinct Architects Directory from British magazine "Wallpaper" and also to be part of the Emerging Architects of 2010 from American magazine Architectural Record.
Our work has been published in more than 15 countries and took part in both national and international biennales and exhibitions.

 ## Francois Frossard

Francois opened his own design firm FFD In 1998 in Miami. He pioneered a new era in nightlife design. He set new standards in interior design around the world.
Francois confronted each venue with a careful meditation on style, function and originality; a marriage of art and business with a profound understanding of his extremely fickle clientele. Francois deals with the complexities of building multiple venues in various cities at the same time masterfully, considerate of local building codes, politics, labor laws and community; designing showpieces capable of withstanding time. Respectfully, Francois cherishes each day as a gift and embraces tomorrow's mystery. He does not take his good fortune for granted and applauds his team, talented craftsmen, manufacturers, designers and tradesmen that service his projects state-wide and internationally. As his company expands further into world class markets his grounded classically trained past narrate his future.

gt2P

gt2P, parametric design and digital fabrication studio.
gt2P was born when Alexy, Guillermo, Tamara and Sebastian found that they had shared interests in the productive processes of architecture and design, such as manufacture, the manufacture or construction, and the possibility of generating them with digital media.

Gulla Jonsdottir

Critically Acclaimed Designer, Gulla Jonsdottir, a 2009 recipient of Hospitality Design's "Wave of the Future" award, is the creative and visionary force behind some of the most striking hotels, restaurants, nightclubs, and spas in the world. As founder of her new design firm, G+ Gulla Jonsdottir Design, located at La Peer Drive and Melrose Ave., Jonsdottir offers her clients a myriad of services including design, architecture, lighting, interiors, and furniture.
Icelandic-born Jonsdottir, spent the eight years as Vice President and Principal Designer of the acclaimed Dodd Mitchell Design (DMD) firm in Hollywood where she was responsible for envisioning and designing numerous projects including the Hollywood Roosevelt Hotel, Cabo Azul Resort in Los Cabos, Mexico, The Thompson hotel in Beverly Hills among many others.
Prior to joining DMD, Jonsdottir spent four years at the renowned Richard Meier and Partners as part of the design teams for projects such as the Getty Center in Los Angeles and the Gagosian Gallery in Beverly Hills. She also gained valuable experience as a Set Designer at Walt Disney Imagineering where she collaborated on Tokyo Disney Seas and Euro Disney in Paris, France.

Gwenael Nicolas

Gwenael Nicolas describes himself as a "choreographer of space" as his work concentrates on "the art of encounter". Knowing is not necessarily understanding. Information can be shared but an experience needs to be encountered to be fully conprehended. Therefore people should be more curious and see, touch and smell by themselves. After impressionism, minimalism, modernism and other "ism"s, now we enter the age of "experientialism".
His designs are well known as translucent, with emotional coloring and attractive forms. When people use his products or encounter his spaces, they suddenly realize that his designs are not only about obvious aesthetics but the fact that considerable thought has also gone into their functionality. The designs originate from a storyboard with subjects as the central focus to which he always incorporates an element of discovery and unpredictability.

 ### Hernandez Silva Arquitectos

Office formed by Mexican architect Jorge Luis Hernandez Silva in 1988, Hernandez Silva Architects works with a group of qualified professionals, which has been highlighted by the quality of architecture produced in both residential, commercial and residential interiors as the buildings.

 ### ID & Design International

ID & Design International is a world-class diversified team of talented designers, architects, and branding consultants with extensive experience specializing in commercial mixed-use, retail, hospitality, and entertainment projects with vast international experience and exposure to some of the world's leading retailers, developers and investors. The IDDI team collaboratively leads and takes a logistical business approach to all design solutions, and transforms them into uniquely branded environments.
As President & Creative Director of IDDI, Sherif Ayad's distinctive style and passion for design has been recognized around the world in a multitude of trade publications, books, and awards throughout the years, contributing to his recognition as one of the leaders in the design industry since 1979.
IDDI is one of the most respected design firms in the world and is known for providing compelling design solutions that build strong brand equity, create market differentiation and ultimately increase sales and productivity.

 ### Ingo Strobel

Ingo Strobel grows up in Frankfurt, Bremen and Berlin. He studied design at Hochschule der Künste in Berlin from 1987 — 1993, and in 1991 he ran his first design office, "designbüro berlin". 1996 he founded motorberlin.
In the field of bars and restaurant-related work he did interiors, installations and consulting for Absolut Vodka, Red Bull, Jägermeister and a couple of other international beverage brands.
His main interest is working in interior and brand design, doing projects for international brands, premium bars and office interiors. He is also booked as a free lance senior designer by agencies, consulting their clients mainly in the field of tradeshow design and brand architecture, and designing such projects. In 2006 in Los Angeles he founded the "Hidden Fortress" with Duardo, an international network of creatives. In 2008 with Hidden Fortress he organised "Made in Berlin", a presentation of young Berlin-based designers at the Salone Internazionale del Mobile in Milano.
Ingo Strobel lives and works with his fiancé Katja and son Noam Oscar Hiendlmayer in Berlin.

Ippolito Fleitz Group

Ippolito Fleitz Group was found in 2002. Managing Partners are Gunter Fleitz & Peter Ippolito.
Gunter Fleitz:
Study of architecture in Stuttgart, Zürich and Bordeaux
worked with Steidle+Partner, München
Project management for the Federal Supreme Court Leipzig for Prof. Stübler
1999 Founding member of zipherspaceworks
Member of Bund Deutscher Architekten BDA
Peter Ippolito:
Study of architecture in Stuttgart and Chicago
worked with Studio Daniel Libeskind, Berlin
Assistant to Prof. Ben Nicholson, Chicago
1999 Founder member of zipherspaceworks
2001.2 Visiting professor at the Academy of Fine Arts Stuttgart
2004–2008 teaching position at the University of Stuttgart
2009 teaching position at the Univeresity of Biberach

 ## Isay Weinfeld

A major name in Brazil, Weinfeld founded his eponymous studio 33 years ago and is renowned for his beautifully simple architecture, often dubbed "Tropical Modern". He is best known for designing stunning villas for various celebrities, but has also turned his meticulous eye to retail projects, nightclubs, hotels, office buildings and furniture.

 ## José Carlos Marques Cruz

José Carlos Marques Cruz was born in Porto in 1957 and graduated in architecture in FAUP, in 1986. He began his work in the same year and continues as a planner till today, sometimes in partnerships with other architects and other times individually. Has already a wide range of completed projects either as architecture or as interior design with the most varied programs, such as single and multifamily housing, shopping and services around Europe, North America, South America and Asia.

 ## karhard architektur + design

Founded in 2002 and based in Berlin.
karhard is an architecture and interior design firm.
The office focuses mainly on the planning and the construction of clubs, restaurants & bars, shops and private housing.

 ## Labics

Labics architecture was founded by Maria Claudia Clemente, Francesco Isidori and Marco Sardella in 2002. The main aim was to convey, under a unique name, those architects, artists and designers who identify with a shared experimentation project in the field of architecture.
Labics research activity aims at an architecture capable of overcoming the status of a singular object, in favor of an architecture that becomes territory, background, structure.
This territorial and structured architecture of Labics detects settlement principles capable of establishing dialectic relationships with the surrounding environment and building a field rich in possibilities, a public and shared space. From 2002 to 2008 Labics has accomplished many professional assignments.

 ## Lionel Ohayon & Siobhan Barry

Over the last 8 years, Lionel, has grown ICRAVE from a two-person operation into a 27-employee, multi-million dollar business. A Canadian-born designer, Ohayon graduated from the University of Waterloo School of Architecture in 1994.
Siobhan Barry is a Partner and the Studio Director at ICRAVE and has been with the firm since its beginning. She hails from Toronto Island, and she graduated in 1997 and achieved the highest award (RAIC medal) for her graduating thesis in architecture.
She moved to New York after graduating, and established a career in retail/boutique design and high-end residential. She then worked for MSM Architects (now MacKay Architecture/Design).
In 2002, Siobhan joined Lionel Ohayon to launch ICRAVE, a start-up design/build firm specializing in hospitality projects.

 ## LOFF

LOFF is a Lighting Design and Architecture Office, founded in 2009 by Cláudia Costa, with the intent to create a dynamic approach to architecture by combining it with Light, a fundamental element for creation, as well as of human well-being. The ultimate goal is to show the clients that by demanding quality, one can aspire to better spaces, better light, and better life. LOFF wish to explore a constant and creative way to exchange knowledge and experiences between many different subjects. LOFF aspires to be a Lighting Design and Architecture Office with a global approach, taking advantage of its privileged situation regarding contacts and partnerships, and at the same time, making the client satisfaction a priority, by investing in rigor and quality as fundamental values for success.

love – architecture & urbanism

love – architecture & urbanism was founded in Graz (Austria) in 1997 and is being managed by three associates (Mark Jenewein, Herwig Kleinhapl, Bernhard Schönherr). Since its foundation, the team is heavily engaged in developing and realizing intelligent and innovative solutions for architectural and urban concepts.

The team has already realized projects in Austria, Japan, USA, Korea and Germany, and also won a number of competitions. They has been awarded several prizes and has been invited to take part in international competitions as well as Exhibitions and lectures.

Since 2003 the brandfield – built identity development company is being managed by the same owners. brandfield develops brand, marketing, experimental and identity strategies for real estates.

 ## Naço Architectures

Naço is a French Architects Studio based in Paris, Shanghai and Buenos Aires, known for its creative & innovative design solutions. The founder Marcelo Joulia set up the agency in Paris in 1991, now the dedicated global team includes 38 designers in architects, interior & graphic design.In October 2005, Naço set up a design studio in Shanghai. The studio is located in Bridge 8 (phase 1) within the old French Concession. Naço Shanghai is a young & dynamic design agency with a total of 22 designers including architects, interior designers, product designers and graphic designers.Their recent projects on hotels and retail spaces in China achieved great results for both the clients and studio itself.
www.naco.net

 ## Planet 3 Studios

Planet 3 Studios, a young, internationally award winning practice represents the vanguard of future-forward design in India. The essence of our way is to address the fundamentals through intense programming, offer solutions that balance wit and wisdom, keep it fresh and never lose the visual appeal. The focus is always on the context, the constraints and the opportunity that a project presents. We like to create designs that maximise the positive impact of design on the environment, use appropriate technologies & materials and source labour locally.

 ## Robert Majkut

Robert Majkut is one of the most important Polish designers. His hard-earned brand and consistency in his approach to design makes him recognizable as a very popular designer and creator of unique places.
As a designer he is sensitive to forms which are humanoid, oblong, round, close to nature. It is a recognizable motif in his works. He looks for inspirations in the surrounding reality. Design in his understanding is based on pioneer, progressive solutions, changing our reality for a specified, meaningful, good purpose. For a long time he has been faithful to the message: "the future is tomorrow invented today".

 ## Shahira H. Fahmy

Shahira H. Fahmy , being featured by The Financial Times as the designer who marks Egypt's debut at the show in Milan in 2007, her work has been acclaimed internationally and her designs have been credited with a number of prestigious architectural and design awards.

Shahira studied architecture at the Faculty of Engineering, Cairo University, Egypt, graduating with honors in 1997, where she also obtained her Masters degree in architecture, in 2004.

In 2005, Shahira H. Fahmy Architects has emerged as one of the most innovative design-focused practices in the middle-east. Her projects — which range from large scale urban and housing projects to interiors and product design — include the Performing and Visual Arts Theater and the Oriental Hall in the American university of Cairo, the Designopolis urban space and many housing projects assigned by SODIC Real Estate. This eco-friendly and culturally aware urban piece has been selected in a number of international awards: recipient the Green Good Design Awards 2010 (Chicago Athenaeum), commended in the Architectural Review MIPIM Future Project Awards 2009, and shortlisted in the World Architecture Festival (WAF) 2009 and Dubai Cityscape 2009.

Shaun Clarkson

Shaun Clarkson has been closely involved with the UK restaurant, bar and nightclub industry for over twenty years and has forged an exciting career since leaving college with a degree in fine art.

With a penchant for the avant-garde and the extravagant, he first cut his teeth with Oliver Peyton's RAW Club in the 1990s where he was resident designer for five years. His relationship with Oliver continued with the refurbishment of the legendary Atlantic Bar & Grill and most recently the Wallace Restaurant at the Wallace Collection. Shaun's other work in the 1990s also included some of London's most iconic and original style bars including Pop, Denim and 10 Rooms – venues that shaped the bar business as we know it today.

Shaun has also worked extensively with drinks and consumer brands to create striking experiential marketing events; clients have included Absolut vodka, Sony Playstation, Rapido TV, Martell cognac, L'Oreal, Budweiser, Maxxium drinks and Plymouth Gin. Shaun recently created the highly acclaimed Perrier Jouet Champagne Bar for Harvey Nichols that attracted huge attention and a nomination in the Restaurant & Bar Design Awards 2010.

Yasumichi Morita

President of GLAMOROUS co., ltd.
Born in Osaka, Japan in 1967. Starting with the project at Hong Kong, China in 2001, his work is successfully and expending globally to cities including New York, London and Shanghai. His creative activities have expanded into graphic and product design beyond his original career in interior design. Prize; THE LONDON LIFESTYLE AWARDS 2010(AQUA LONDON), The Andrew Martin Interior Designers of the Year Awards etc…

Yusaku Kaneshiro

1960 Born in Okinawa.
1979 Graduate from Okinawa Prefectural Koza High School.
1981 Graduate from Tokyo Designer Gakuin College
1988 Joint established Planning and Analysis for Environment
2000 Established Yusaku Kaneshiro+Zokei-Syudan Co.,Ltd.

Vaillo + Irigaray - Architects

Antonio Vaillo I Daniel, Architect (Barcelona, 1960)
1979 — 1985 Architectural Studies in ETSA University of Navarra.
Juan Luis Irigaray Huarte, Architect (Navarra, 1956)
1974 — 1980 Architectural Studies in ETSA University of Navarra.
Competitions and Prizes (only list some of them)
2010
First Prize, Archizinc Awards, B3 House, Pamplona
First Prize, COAVN Awards, Lounge MS, Cadreita
First Prize, Restricted competition
2009
Gold Medal, Miami Beach 2009 Bienal, Interior Design, D Jewellery Shop, Pamplona
First Prize, Restricted competition
First Prize, Interior FAD'09 Opinion Award, Elmerca'o Restaurant, Pamplona
TOP 50 Interior of Spain 2009, Convened by VIA Group
2008
First Prize, Restricted competition
First Prize, Interior FAD'08

后记

本书的编写离不开各位设计师和摄影师的帮助,正是有了他们专业而负责的工作态度,才有了本书的顺利出版。参与本书的编写人员有:

Jorge Luis Hernández Silva, Carlos Díaz Corona, Andy Martin, Vangelis Paterakis, John Horner, Kyna Leski, Jack Ryan, Nick Croft, Aaron Brode, Eleanor Lee, Olga Mesa, Manuel Cordero, Chris Bardt, Yasumichi Morita, Nemu Qiang, Nacasa, Gwenael Nicolas, Antonio Vaillo, Juan Luis Irigaray, Daniel Galar Irurre, Kalhan Mattoo, Santha Gour Mattoo, Hina Chudasama, Prashanta Ghosh, Gauri Argade, Mrigank Sharma, India Sutra, Thomas DARIEL, Benoit ARFEUILLERE, Francesco Gatti, Daniele Mattioli, Abraham Cherem Cherem, Javier Serrano, Jaime Navarro, KAN TAI LAI, Robert Majkut, Cláudia Costa, haun Clarkson , Baldauf, Cattonuliana Nohara, Flavio Faggionernando Forte, Lourenço Gimenes, Rodrigo Marcondes, Ferraz, Bruno Milan, Carolina Matsumoto, Castilha Iluminação, Foz Engenharia, Fran Parente, Gulla Jonsdottir, Ryan Forbes, Antonio Di Oronzo, Lionel Ohayon, Siobhan Barry Frank Oudeman, Alan Barry, Shahira H. Fahmy, Sankie. L, Rob Wagemans, Janpaul Scholtmeijer, Charlotte van Mill, Erik van Dillen, Isay Weinfeld, Colorfuldust, Gian Frey, James Dyer-Smith, Dimitris Naoumis, Andrea Ottaviani, Miguel Rodenas, Jesús Olivares, Lionel Ohayon, Siobhan Barry , Francois Frossard, Jason Lane, Barbara Rourke, Jason St. John, Naço Architectures, Ingo Strobel, Winka Dubbeldam, Axel Schaefer, Rob Wagemans, Joris Angevaare, Erik van Dillen, Rob Wagemans, Jeroen Vester, Ulrike Lehner, Erik van Dillen, Melanie Knüwer, Jari van Lieshout, Sonja Wirl, Nina Schweitzer, Peter Ippolito, Gunter Fleitz, Tim Lessmann, Hakan, Sakarya, Yuan Peng, José Carlos Marques Cruz, Yusaku Kaneshiro, Cass Calder Smith, Bryan Southwick, Barbara Turpin-Vickroy, Djordje Gec.

 ACKNOWLEDGEMENTS

We would like to thank everyone involved in the production of this book, especially all the artists, designers, architects and photographers for their kind permission to publish their works. We are also very grateful to many other people whose names do not appear on the credits but who provided assistance and support. We highly appreciate the contribution of images, ideas, and concepts and thank them for allowing their creativity to be shared with readers around the world.